STUDENT LIBRARY
30106 006718257
AF750246

MEAT PRESERVATION

PREVENTING LOSSES AND ASSURING SAFETY

F N P

PUBLICATIONS IN FOOD SCIENCE AND NUTRITION

Journals

JOURNAL OF FOOD LIPIDS, F. Shahidi
JOURNAL OF RAPID METHODS AND AUTOMATION IN MICROBIOLOGY, D.Y.C. Fung and M.C. Goldschmidt
JOURNAL OF MUSCLE FOODS, N.G. Marriott and G.J. Flick, Jr.
JOURNAL OF SENSORY STUDIES, M.C. Gacula, Jr.
JOURNAL OF FOODSERVICE SYSTEMS, C.A. Sawyer
JOURNAL OF FOOD BIOCHEMISTRY, J.R. Whitaker, N.F. Haard and H. Swaisgood
JOURNAL OF FOOD PROCESS ENGINEERING, D.R. Heldman and R.P. Singh
JOURNAL OF FOOD PROCESSING AND PRESERVATION, D.B. Lund
JOURNAL OF FOOD QUALITY, J.J. Powers
JOURNAL OF FOOD SAFETY, T.J. Montville
JOURNAL OF TEXTURE STUDIES, M.C. Bourne

Books

MEAT PRESERVATION: PREVENTING LOSSES AND ASSURING SAFETY, R.G. Cassens
S.C. PRESCOTT, M.I.T. DEAN AND PIONEER FOOD TECHNOLOGIST, S.A. Goldblith
FOOD CONCEPTS AND PRODUCTS: JUST-IN-TIME DEVELOPMENT, H.R. Moskowitz
MICROWAVE FOODS: NEW PRODUCT DEVELOPMENT, R.V. Decareau
DESIGN AND ANALYSIS OF SENSORY OPTIMIZATION, M.C. Gacula, Jr.
NUTRIENT ADDITIONS TO FOOD, J.C. Bauernfeind and P.A. Lachance
NITRITE-CURED MEAT, R.G. Cassens
POTENTIAL FOR NUTRITIONAL MODULATION OF AGING, D.K. Ingram *et al.*
CONTROLLED/MODIFIED ATMOSPHERE/VACUUM PACKAGING OF FOODS, A.L. Brody
NUTRITIONAL STATUS ASSESSMENT OF THE INDIVIDUAL, G.E. Livingston
QUALITY ASSURANCE OF FOODS, J.E. Stauffer
THE SCIENCE OF MEAT AND MEAT PRODUCTS, 3RD ED., J.F. Price and B.S. Schweigert
HANDBOOK OF FOOD COLORANT PATENTS, F.J. Francis
ROLE OF CHEMISTRY IN THE QUALITY OF PROCESSED FOODS, O.R. Fennema, W.H. Chang and C.Y. Lii
NEW DIRECTIONS FOR PRODUCT TESTING OF FOODS, H.R. Moskowitz
PRODUCT TESTING AND SENSORY EVALUATION OF FOODS, H.R. Moskowitz
ENVIRONMENTAL ASPECTS OF CANCER: ROLE OF MACRO AND MICRO COMPONENTS OF FOODS, E.L. Wynder *et al.*
FOOD PRODUCT DEVELOPMENT IN IMPLEMENTING DIETARY GUIDELINES, G.E. Livingston, R.J. Moshy, and C.M. Chang
SHELF-LIFE DATING OF FOODS, T.P. Labuza
ANTINUTRIENTS AND NATURAL TOXICANTS IN FOOD, R.L. Ory
UTILIZATION OF PROTEIN RESOURCES, D.W. Stanley *et al.*
VITAMIN B_6: METABOLISM AND ROLE IN GROWTH, G.P. Tryfiates
POSTHARVEST BIOLOGY AND BIOTECHNOLOGY, H.O. Hultin and M. Milner

Newsletters

MICROWAVES AND FOOD, R.V. Decareau
FOOD INDUSTRY REPORT, G.C. Melson
FOOD, NUTRITION AND HEALTH, P.A. Lachance and M.C. Fisher
FOOD PACKAGING AND LABELING, S. Sacharow

MEAT PRESERVATION

PREVENTING LOSSES AND ASSURING SAFETY

by

Robert G. Cassens, Ph.D.

DEPARTMENT OF MEAT AND ANIMAL SCIENCES
UNIVERSITY OF WISCONSIN
MADISON, WISCONSIN 53706

FOOD & NUTRITION PRESS, INC.
TRUMBULL, CONNECTICUT 06611 USA

Copyright © 1994 by

FOOD & NUTRITION PRESS, INC.

Trumbull, Connecticut 06611 USA

All rights reserved. No part of this publication may be reproduced, stored in a retrieval system or transmitted in any form or by any means: electronic, electrostatic, magnetic tape, mechanical, photocopying, recording or otherwise, without permission in writing from the publisher.

6718257

LEEDS UNIVERSITY LIBRARY

Library of Congress Catalog Card Number: 94-70368

ISBN: 0-917678-34-6

Printed in the United States of America

ACKNOWLEDGMENTS

Sincere appreciation is expressed to the University of Connecticut for awarding me with a Short-Term Guest Professorship. It allowed me to reside in Storrs from January through March 1992, and to think about and work on the manuscript that eventually became this book. Cameron Faustman was responsible for initiating the award, for many thoughtful discussions and for making sure I was provided for and comfortable. Thanks is expressed to all faculty, staff and students of the University of Connecticut's Department of Animal Science for making me welcome and for integrating me into the daily activities. I identify specifically Don Kinsman who was a steady source of information about Meat Science and Connecticut history, and Lonnie Malkus who cheerfully shared his office and put up with my constant pecking at the keyboard. Also, I thank the University of Wisconsin for providing me with more than 30 years of gainful employment and for encouraging projects such as the one described above.

Several individuals have read all or parts of the manuscript and provided me with numerous important and useful suggestions. I thank them and list their names as follows: several anonymous reviewers, L. Borchert, J. Johnson, D. Kinsman and C. Faustman.

I thank John J. O'Neil and Christine L. Murphy of Food & Nutrition Press, Inc., for working with me during final preparation of the manuscript and for providing expert editorial assistance. I thank Sue Ann Hubanks for her professional proofreading, spell and grammar checking and word processing services.

PREFACE

This book is about the preservation of meat. It is written as an integrated and all-encompassing text that includes: historical aspects and trends, basic background information, the evolution and status of techniques and procedures, and potential future developments. The latter are especially important because there is a definite change occurring, based on consumer desires, to produce and market meat and meat products that have been subjected to a lower degree of preservation, yet appear to be fresh and more healthful. There is an intense interest in producing the safest meat possible. The overriding theme I am presenting is how to understand the science of meat and employ known technologies to accomplish that goal.

My aim was to write a readable and informative account—not a fact-laden reference text. It represents my thoughts and efforts. Obviously, what I have written has been influenced by contact and consultation with numerous colleagues. The topic is indeed important. Without proper preservation of meat, there is a certainty of economic loss due to spoilage and the real danger of transmission of food borne illness. Most information about preservation of meat is found strewn throughout the scientific literature. I have attempted to draw all aspects together and weave into the writing an analysis of underlying reasons for what has happened—as well as what might happen. References are not cited directly, but as general sources. I did this to make the text flow more smoothly while at the same time providing a resource for those wanting further information.

I tried hard to give an accurate, objective and full description without inserting my own biases or opinions. This proved difficult at times when I was tempted to counter the wave of ungrounded accusations presently being directed against meat.

I intended the book primarily for undergraduates but believe it will be useful reading for anyone involved with or wishing to learn about meat preservation.

ROBERT G. CASSENS

CONTENTS

CHAPTER 1

INTRODUCTION

Obtaining and consuming food is fundamental to survival. During prehistoric time, hunters and gatherers devoted most of their available time to the task. And, they were amply rewarded by providing themselves and their families a varied and nutritious diet that obviously sustained them in their active and demanding pursuits—often under harsh environmental conditions.

Now, we deal with food several times a day, and it is still prerequisite to our survival. The human body has complex nutritional requirements that must be fulfilled. However, it now seems that the major concerns are about choices and desires, within the bounds of what we are able and willing to pay. There is an amazing variety from which to choose. We consider infrequently the availability of food, and even less frequently the fact that it is necessary for our existence.

Many factors impinge on our choices about food. The senses of satisfaction and enjoyment influence what foods we choose and how much we consume. Food choices play a major role in various social settings—these range from day-to-day life at the family table, through events planned solely for entertainment, to official or business meals at which rather serious decisions may be made or influenced.

Convenience is becoming more and more of a consideration for some. Cultural and religious factors may dominate certain choices about food as may habits ingrained during preceding generations. And finally, a good deal of marketing effort is aimed at influencing our choice of foods—and, it often triumphs.

As food choices are made, there is a somewhat vague sense of how the food will affect our health and well-being—both from the positive viewpoint of providing required nutrients and from the negative viewpoint of potential food safety problems. The latter concern has been fueled by activist groups promoting special interests and by coverage in the media. The activist groups, usually located in Washington and well-funded, have made a career, since the late 1960's, of attempting to affect food policy by the legal-political and public relations route while ignoring scientific fact.

An interesting contrast exists between times of plenty and times of need. During the Great Depression, and the World Wars, food was often difficult to find and was even rationed; so the major concern was just obtaining it. When food is abundant, people make more choices about type and quality, and are even more inclined to complain about it.

Still, the basic reason we must have a continuing, adequate and varied supply of food is to keep our bodies functioning. Without food in proper amount and appropriate balance we would cease to exist. Therefore, the preservation of food is of momentous consequence simply because preservation makes it more available. Preservation makes possible conservation of surplus foods that might otherwise be wasted if not consumed fresh.

Preserving techniques have a monetary impact by lowering the loss associated with spoilage of perishable foods. A complicated distribution network is permitted. In addition, preserving techniques protect against a wide array of foodborne illnesses.

This book is about preservation of food—and specifically about preservation of meat. The presentation is made by intertwining three viewpoints. The scientific understandings of various preserving techniques are explained; the historical developments and present situation are considered; and an analysis is made, of both past and potential accomplishments.

EARLY ORIGIN OF MEAT PRESERVATION

Meat provides a special case for study, as it was the common objective for early hunters and was much desired. The appetite for meat probably came from a sense that it was in some way necessary for bodily well-being. Some cultures or nations evolved with a great preference for meat-based diets, which still continues to the present.

Because of its composition (wide variety of essential nutrients and high water content), meat is not only a highly nutritious food, but also is an extremely perishable product compared, for example, to some plant products such as grains. After a successful hunt, there was often more than could be consumed; but if it was not eaten in a short time it spoiled and another hunt had to be launched. In order to have a more or less continuous supply, there was an ongoing concern for ways to keep meat in an edible state, if only for a few additional days. In addition, spoiled meat may cause human illness.

The story really begins in the Stone Age. This period commenced about 2.5 million years ago, and ceased in the region of Mesopotamia and Egypt about 3,000 B.C. when the use of bronze was beginning. The Old Stone Age or Paleolithic people depended entirely on hunting, and this period comes up to about 8000 B.C. The most recent or Neolithic phase led to the development of farming with domesticated cereals and livestock.

The richest and earliest farming, which included for example, irrigation of crops, was present about 5,000 years ago in the Fertile Crescent. This great civilization developed in the crescent shaped area running from the

Mediterranean Sea, stretching along the Tigris and Euphrates Rivers, and ending at the Persian Gulf.

Different periods of the Stone Age were occurring in various areas of the world simultaneously. The development spread into Europe and during the Bronze Age it appears that foods were basic and adequate and included meat, eggs, fish, cereals and some fruits and vegetables in season. The ideas of heating and cooling, drying, fermenting, salting and packing in oil or other substances to exclude air were in hand.

In general, prehistoric man existed merely by gathering what was available. Eating depended upon hunting, fishing and picking what could be found. As modern civilization began to take shape during the period 10,000 B.C. to 3,000 B.C., the attitude of making better and more reliable provisions for food began to emerge. Some began to keep animals and to plant seeds.

The first genuine attempts to preserve food also emerged. Heating and drying were employed, as well as cooling. Salting and drying were practiced in the Nile valley. Dried grains were held in storage to guard against a potential shortfall. Fermentation was used as a means to produce alcoholic beverages. In warm climates it became known that the best way to keep perishable foods was to transform them to altered products by allowing or encouraging fermentation. In climates with change of season, advantage was taken of the cooler weather to keep fresh meat longer.

The Greek and Roman civilizations were advanced and organized in regard to preparing a diversified assortment of foods. Meat, in the form of sausages obviously played an important role, and methods of preservation, such as salting and pickling, were well-established.

During this early period of time, the sequence was from hunting and gathering to pastoral life, to more or less settled farming with cultivated crops and domesticated livestock. Villages and cities were formed, signaling the beginning of the need for transportation. Various preparation methods began to surface. The need for food preservation was recognized, and many interesting and useful procedures were developed. Substantial improvements were made and these were done by trial and error—and formed the basis of much of what we still practice today.

ESTABLISHMENT OF THE MEAT INDUSTRY IN AMERICA

Our story now jumps ahead to the United States during the 1600's. The norm then for fresh meat was to raise animals for home consumption, and to use to the extent possible seasonal changes to take advantage of cooler temperatures for slaughter of animals. Alternatively, the animals could be moved on foot to centers of population and then slaughtered and distributed quickly.

The practice of "packing pork" illustrates a specific preservation procedure. The meat was cut into pieces and packed in barrels with generous amounts of salt, and perhaps other ingredients. This pork could be stored for subsequent use, or sometimes it could be exported to other countries and sold at a profit. This practice was actually the genesis of the meat packing industry in the United States. William Pynchon began packing pork in Springfield, Massachusetts, about 1660.

The need for food preservation became especially evident beginning about 1800 as industrialization began to occur. Massive numbers of people began living in cities where there were jobs. But, the food was produced in the rural areas. The first major problem was to transport sufficient food to the people, and preservation methods would not only simplify transportation, but also allow the city dwellers to keep stores of food on hand.

The technique for canning was published by Appert in 1810. The process worked well, but the resulting preserved foods were quite expensive. They were used in the first instance by the military, or very wealthy people, or on expeditions. It was probably 50 years or more (after the U.S. Civil War) before canned goods were really made available to the general populace. From 1850 to about 1900, there was a large trade in canned meats that came primarily from Australia and South America.

The canning business began in the United States in 1819 when two people immigrated from England. William Underwood came to Boston and Thomas Kensett settled in New York. Both set up canning businesses using the Appert technique, and both prospered. By the 1840's, some quantities of food were being preserved by canning. Interestingly, the patent for canned condensed milk was taken by Gail Borden in 1856. Libby, McNeil and Libby put canned corned beef on the market in 1872.

The business of packing or preserving meat was influenced greatly by shifts in population centers, by shifts in areas of production, and by transportation considerations. We have noted already that pork was being packed during the 1600's in New England. It was consumed in the area or sometimes shipped to the Southern plantations or to the West Indies.

Following the American Revolution, there was expansion westward with growth in population and establishment of new cities. During the early 1800's, interest was centering in Cincinnati, which was to become a major center for slaughter and packing of pork. By 1860, Chicago emerged as a center of slaughter. Getting the animals to the slaughter centers, which were located in population centers, always involved transporting the animals there—first by driving them on the hoof and subsequently moving them by railroad. The supply of fresh meat for urban areas was obtained by taking the animals there for slaughter. The alternative was to have salted pork that could be transported. By the 1870's canned meat was becoming more generally available.

ADVENT OF REFRIGERATION

About the 1870's, the meat business was revolutionized by the advent of refrigeration. The industry was intensified, and a few of the companies grew to enormous size. Packers were using natural ice by the 1860's, but there were problems associated with harvesting and storing the ice. Plants manufacturing ice with mechanical refrigeration were in place by 1880. The first shipment of fresh meat was made by G.H. Hammond in 1869 and went from Chicago to Boston. A few years later G.F. Swift established a most successful refrigerated meat business. It took a time to carry out proper insulation, and to learn how to position the meat and provide adequate air flow in the cars. The refrigerated rail car was in general use by 1890, and the industry was changed from a small seasonal type of business to an industrial giant.

Another era of meat preservation and refrigeration took place subsequently and was an important aspect of historical development. This occurred during the early 1900's and was associated with shipping fresh beef from producing countries of Argentina and Australia to consuming countries of Europe, and England. Shipment was by sea and either in the chilled or frozen state. There was a considerable amount of scientific research, undertaken and recorded, about engineering, microbiology, use of formaldehyde vapors, use of carbon dioxide, and control of drip when frozen meat was thawed.

An interesting, and not so surprising, point of history regarding these foregoing technological improvements was the fact that competition actually slowed their adoption. For example, as the Chicago packers attempted to use refrigerated cars to sell their beef in the East, there was great resistance from the existing meat people in the East who were concerned that better quality meat would be delivered at a lower price by use of the new technology.

Another less well-known example was the business venture of Marquis de Mores. He attempted a revolutionary idea when he went to North Dakota in 1883 with the purpose of slaughtering beef there and then shipping it back East. He thought slaughtering on the range was preferable to shipping cattle to distant points for slaughter, as advantages could be had in terms of shrinkage, meat quality and a lower price to the consumer. A packing plant was built in Medora, North Dakota, and the Northern Pacific Refrigerator Car Company was established. It became clear by 1886, however, that the business would fail because of opposition from the ice dealers, the railroads, and the meat businesses of Chicago and New York.

These experiences illustrate a common fact. That is that advances in technology are often delayed in adoption by existing business interests that are reluctant to alter their procedures, or realize they cannot compete effectively with the new and improved product.

A major achievement was the development of procedures in the 1920's by Clarence Birdseye, which led to the commercialization of frozen foods. Frozen food lockers were used extensively during the 1930's and continued as active businesses during World War II and the remaining 1940's. They were being phased out during the 1950's as home, chest freezers were becoming more available and popular. The frozen food lockers were associated closely with the meat industry, as many were operated with a custom slaughtering and meat processing business.

CHANGES IN THE INDUSTRY

Before describing later changes in the meat business, as related to preservation, let's examine briefly the before and after of major changes that occurred around the 1860's. A physical movement or urbanization of people was underway. The Civil War made evident the great need for food and the intricacies associated with preserving and transporting it. Production areas were removed from the major consumption areas. As the farming business was being shifted westward to the Ohio River Valley, the animals produced were still being driven back to slaughter and consumption centers, such as Baltimore and Philadelphia. Refrigeration was unknown, except for natural ice, and so the meat business was quite seasonal. After the Civil War, the advent of mechanical refrigeration with development of the railroad system allowed a great change in the industry. Distribution of an adequate and year-round supply of fresh meat to the population was permitted.

The Union Stock Yards was opened in Chicago in December 1865, and subsequently, Chicago became the largest meat packing center in the nation. The major meat packers flourished. Animals were moved in immense numbers to Chicago for slaughter, and carcasses were shipped out to feed the nation. An important series of events happened in the early 1900's. Upton Sinclair used his book *The Jungle* to describe intolerable working conditions and unsanitary practices in the packing plants. As a direct result, food laws and governmental regulations were put in place to ensure the production and distribution of the highest quality and safest meat possible. Details of governmental regulatory programs and efforts by the industry to improve the product are explained in subsequent chapters.

After World War II, other major changes were underway and by the 1950's and 1960's, events were occurring that would result in closing of the centralized market and slaughter concentration located in Chicago. The industry was beset with large price fluctuations, and difficulties with labor costs were a continuing problem. During the 1950's and 1960's, freezing of raw material (meat) became a common practice as a means to control inventory and to deal with market fluctuations. The interstate highway system evolved and trucks began to replace the rail system. Large feedlots were established in the west and modernized packing plants moved further west.

Instead of whole carcasses being transported, wholesale cuts were shipped, and often in vacuum bags. This method of fabricating and transporting became known by the term "boxed beef". It eliminated shipment of whole carcasses and provided a good environment, in the bag, for controlled aging of the meat. This system continues in use today and is extremely efficient at providing consumers in population centers with fresh beef processed in distant production areas.

Another change has been the separation of slaughter and processing. Classically, a slaughter plant slaughtered animals, and then sold not only fresh meat but also processed some of it into various products. Now, certain companies concentrate only on high volume and efficient slaughter of animals and fabrication of carcasses while other companies specialize solely in further processing of the fresh meat they purchase.

All these significant changes in the industry brought their own special problems, and opportunities, regarding the practice of meat preservation.

Before 1800, there was a rudimentary understanding of some preservation techniques, and procedures such as cooling, heating, pickling, fermenting, drying and salting were knowingly used. After 1800 a genuine understanding of the needs and means of preservation began to develop, and some enormous advances, such as preservation by canning, were made. Many of these advances in food preservation techniques can be related directly to military considerations. Transportation was another important influence, and now, I believe, food preservation is being influenced strongly and directly by the attitudes of the consumer—especially as related to environmental and societal issues.

INFLUENCE OF THE MILITARY

The role of a strong military force in shaping history is evident. In order to maintain an effective military effort, the troops must be fed well. This means the food must be preserved so that it can be transported to the troops, and it must be of sufficient quality when it arrives that they will consume it. Nothing can be more devastating to a military effort than to have: (1) no food, (2) food in such a tainted or spoiled state that it is rejected, or (3) crippling sickness due to a foodborne illness.

Meat has always played a central role in military feeding. This is because of nutritional considerations as well as the phenomenon of morale.

The best known example of the influence of the military on food preservation was the stated desire of Napoleon and the French Army to have an adequate method of food preservation—a situation that indeed stimulated Nicholas Appert to devise the process we know today as canning.

Similarly, the Navy has had major concern about providing adequate food during long voyages. The early disasters associated with development of nutritional problems on board ship were well known, but the cause and cure were not understood. Subsequently, it was learned that having fresh or properly preserved foods would prevent nutritional maladies.

Food was a key element during the Civil War with the Union being well provisioned and the Confederates being less better off, especially as supply lines and production areas were effected. Packed pork (salted in barrels) was the preserved food of choice, and use was also made of canned vegetables and fruits. The needs during the Civil War for meat resulted in a tremendous market for the meat packers and, in fact, established certain of the companies as future industrial giants.

Immense use of canned goods was made during the World Wars. Many of the more modern techniques of food preservation are associated with World War II, and especially from research by the U.S. Army Food and Container Institute, located formerly in Chicago, and subsequently by the U. S. Army Natick RD & E Center located now in Natick, Massachusetts. Examples of these include heat processing, freeze-drying, irradiation and intermediate moisture foods. This work is ongoing and has taken into consideration the needs of the space program. In the Persian Gulf Conflict, food again played an important role as evidenced by the press descriptions of MRE's (Meals Ready to Eat). The MRE's are basically a new generation of preserved and ready to eat foods. But, it has surfaced already that a major problem with these rations was the question of how to deal, in an environmentally responsible manner, with the enormous amount of packaging material generated in relation to the small amount of food. The military continues to be a leader in the search for better methods of food preservation.

ROLE OF TRANSPORTATION

The developments in transportation and food preservation techniques have been closely interrelated. There are two aspects. One is to provide a source of food to those doing the traveling, and the other is to devise an effective transportation to move a food supply from one location to another.

A much desired insurance to any long journey is to include a preserved and transportable food for use not only daily, but also in emergency situations. All travelers are faced with feeding themselves. This basic need obviously stimulated early searches for preservation techniques, and as we have already discussed, the need became more critical, for example, with long ocean voyages (with little or no prospect for fresh food for months at a time) and lengthy military campaigns.

On the other hand, as some forms of transportation developed—railroads, for example—the movement of food from producing to consuming areas could be much more easily accommodated. Still, other preservation problems had to be solved. For example, fresh meat could be transported by rail. Before this could be done, however, refrigeration had to be incorporated or else it would have remained the better choice to transport the live animals to the population centers for slaughter.

With air travel, special problems of preserving pre- or partially-cooked foods for use during flying time had to be solved. Also, airplanes permitted rapid transport of fresh items from distant places—at a cost.

While the space program seems to present special problems related to food, they are in fact little different from those faced by early explorers. It is a matter of how to carry an adequate supply of food that will not spoil during the journey.

INFLUENCE OF THE CONSUMER

Two major industries—the military and transportation—have influenced food preservation techniques, and they will probably continue to do so. The present era may be known as that of consumer influence. The consumers, via their expressed demands, are having major impact on use of old and development of new techniques for preservation of food.

This situation presents some most interesting aspects that are resulting, almost, in an impasse. To the objective observer it must appear as a rather unfortunate dilemma. Consumer demands are growing for fresh-like, preservative-free, stable, convenient-to-use, healthful foods that have been produced and delivered with a minimum usage of energy and result in a minimum of solid waste to send to the landfill. Most efforts to satisfy these demands are expensive processes that employ intense preservation techniques and extensive packaging.

Many people of the world have a variable and more than adequate supply of food at a very nominal cost. These people do have a growing concern about the safety of the food and about the environmental and ecological cost of producing, processing and transporting it. However, another segment of the world's population does not have adequate food, and they have little or no opportunity to use current available information about production and preservation techniques. Is there an answer or middle ground? I visualize much more political involvement in food policy.

MEAT SOURCES

While some differences exist in production and processing techniques for beef, pork, lamb and poultry, the resulting product, which is meat, from these species is essentially similar and amenable to the same preservation techniques. Throughout the text all meat will be considered together unless specific detail requires comment.

The poultry industry has undergone enormous changes in recent years. Many families kept small numbers of poultry in earlier days. To raise a few chickens was common and easy. They were efficient converters and produced both meat and eggs. It was not until the 1920s that attempts were initiated to raise poultry on a mass production basis. Two major problems facing the large scale production attempts were how to provide a completely balanced ration to the birds in confinement and how to control disease. Research solved these problems, and following World War II there was a steady growth of the industry. By the 1960's, the industry really began to expand to the point where there were a few dominant companies and most of the birds were produced under contract.

Also consumers began to demand more poultry meat—because it was less expensive and because there was a perceived health benefit to the "light" poultry meat. A large market was also developed for processed poultry products. This business was opened quickly because the processing and preservation techniques, which had been developed originally for red meat, were merely then applied to poultry meat. Therefore, development time and cost were minimal.

The meat produced from fish muscle can also be preserved by all the classical methods. However, there are some differences, and fish is not discussed directly in this text. Also, because of regulatory interest in the fish industry there may be significant changes occurring soon. A major difference is that most fish are harvested or captured away from land so preservation must begin or take place on board the ship. However, more and more fish are being produced by culture techniques. In general, fish is more perishable than other muscle foods, meaning that preservation must be done rapidly and efficiently. The fact that the bacteria usually associated with fish are predominantly psychrophiles can also result in special problems, especially if cross-mixing with typical red meat occurs. Processing fish into surimi has become a popular way to use it and convert it to a value-added product.

There is an extensive and interesting history associated with the early search by humans for meat and with the subsequent development of the modern meat producing industry. Preservation has been a factor important during all phases of the history. Societal considerations, conquest at local and global levels, and technological developments have influenced the subject of meat preservation—and will continue to do so.

CHAPTER 2

UNDERSTANDING MUSCLE AND MEAT

Meat is defined broadly as the flesh of animals used for food. In a more strict sense, meat is composed chiefly of muscle. This means that bone has been removed and the surface of the muscle is devoid of or at least relatively free from fat. Although some classical cuts of meat such at T-bone steak, contain bone and others may contain substantial surface or seam fat, the convention used is that meat is considered to be muscle.

What are consumer perceptions of meat? When the consumer steps up to the retail meat counter to contemplate a purchase decision there are three factors being weighed. These are appeal, wholesomeness and price.

If the meat does not provide a sensation of appeal it will not be purchased. Undoubtedly, the most important perceptions in this regard are freshness and quality—how the consumer judges these considerations is difficult to define or explain.

Color is the first characteristic noticed by the consumer and is the factor most used to judge both freshness and quality. A bright, uniform attractive color is related to freshness. If the surface color appears dry, or if off-color areas are present the meat is judged as not fresh. Color is also used as an indicator of species and to make a judgment whether the meat came from a young or old animal. Beef has more myoglobin (appears redder) than lamb, which in turn has more than pork. Poultry is lower still. As an animal matures, the myoglobin content of the muscle increases several fold.

Quality is more difficult to define. Certainly the foregoing factors are involved, but also such aspects as texture, firmness and composition are considered.

At present, external fat on the surface of meat is trimmed to a minimum at retail because of consumer demand for low-fat foods. Debate continues as to the role of intramuscular fat (known as marbling) in quality. The USDA quality grades take major account of intramuscular fat or marbling. While some have dismissed marbling as a factor in palatability (based primarily on tenderness), and there has been a trend to reduce marbling because of concern about dietary intake of fat, others maintain that reducing marbling results in dry, unflavorful meat—a condition that has contributed to decreased consumption of beef.

The aspect of wholesomeness has two facets for consumers. On the positive side, they are aware that meat has certain incontestable nutritional benefits, such as providing required protein (essential amino acids), vitamins and minerals. Satisfaction or satiety is a more nebulous consideration, but consuming a meat-based meal is a fulfilling experience. On the negative side, they are also aware of the concept that meat may have some detrimental affect on health. This perception is greatly influenced by lifestyle and trends expressed in the media. For example, consumers are highly concerned about health as related to saturated fat content of meat and about the possibility of chemical residues such as pesticides, antibiotics and growth stimulants.

Finally, price is a determinant. Consumers have limits beyond which they will not make a purchase.

The foregoing discussion was directed at fresh meat. Application of the same thought pattern can be extended to processed meats (see Chapter 3 for explanation of processed meat). Even though processing normally involves addition of chemicals (such as nitrite and salt), heating and different packaging techniques, the same factors of appeal, wholesomeness and price are still scrutinized by the consumer.

COMPOSITION

In simplest terms, meat is composed of about 20% nitrogenous compounds and about 80% non-nitrogenous compounds. A typical proximate composition is 70% water, 18% protein, 11% lipid, 1% mineral and a trace of carbohydrate. Composition varies enormously—especially the lipid content which in turn affects the amount of water present in the muscle. The reasons for variation in composition and the possibilities for controlling it will be examined in Chapter 3.

Detailed information about composition is provided in Tables 2.1 and 2.2. An estimate of the composition of the typical live animal is given in Table 2.1. Results are expressed as a percentage, and the figures are subdivided into several levels of organization. For example, the total live animal is about 40% non-carcass and about 60% carcass. In turn, that 60% of carcass contains about 60% muscle, 1% tendon and ligaments, 27% fat and 12% skeleton.

A similar presentation is made in Table 2.2 for typical fresh muscle. Composition can be divided easily into nitrogenous and non-nitrogenous categories. Protein is the major component of the nitrogenous category with non-protein nitrogenous compounds making up a small portion of the total. The non-nitrogenous categories include water, lipid, mineral, carbohydrate and vitamins.

Water is the most abundant component of meat, and of interest is the fact that muscle will bind considerably more added water. The water content and the manner in which it is bound influences juiciness, tenderness, color and surface appearance. In the condition of pale, soft, exudative (PSE) pork, for example, the water is loosely bound. As a result, moisture collects on the surface, and because of reflectance properties the muscle appears very pale in color.

In normal meat about one-third of the total water is bound or held tightly, but the majority is more loosely bound and can be squeezed from the muscle by physical force. The so-called water-holding capacity of muscle not only influences physical properties and appearance but also is of definite economic importance. Any loss of water from the muscle, in the form of drip or exudate, means there is less weight to sell.

The pH of muscle influences the water-holding capacity. A plot demonstrating the effect of pH on water-holding capacity is given in Fig. 2.1. Minimum water-holding occurs at pH 5.0 to 5.1 which is the approximate isoelectric point of the major muscle proteins. At the isoelectric point positive and negative charges are essentially equal, and consequently the physical structure is in the most compact or collapsed state affording a minimum of space in which water molecules can reside or be immobilized. As pH is shifted to either side of the isoelectric point there is an excess of either positive or negative charges. Like charges repel, forcing the protein structure apart and providing more space for water to reside.

Muscle fat is composed primarily of neutral triglycerides stored in fat cells, and a lesser quantity occurs as phospholipids located in cell membranes or membranes of cell organelles. Stearic, palmitic and oleic are the predominant fatty acids in animal fat. Oleic, linoleic and linolenic are the commonly found unsaturated fatty acids, and they have one, two and three double bonds, respectively. Pork fat is softer, or more unsaturated, than beef fat because it contains more linoleic acid.

It is essentially impossible to alter the composition of fat by dietary manipulation in the ruminant. In non-ruminants, however, the composition of fat can be changed by altering the diet. The fat in muscle may also develop atypical flavors via dietary factors.

The polyunsaturated fatty acids are considered essential nutrients for humans.

In processed meats, fat is normally present at 20% or more, and it is well known that merely lowering fat content alters textural characteristics and markedly decreases palatability properties. Fat is subject to development of off-flavors through oxidation, with the unsaturated fatty acids being more susceptible. Fat from meat has come under increasing criticism due to its caloric impact in the total diet and potential contribution to obesity, as well as possible relationship to heart disease and cancer. The fat and human health issues have been debated for more than 30 years without a conclusion being reached. The best advice appears to be to follow a moderate course.

A recent suggestion is that vitamins with antioxidant properties (E and C) may play a key role by preventing accumulation of oxidation products of fat. These oxidation products may play a role in heart disease.

TABLE 2.1
LIVE ANIMAL COMPOSITION
By permission of R.G. Kauffman, University of Wisconsin, Madison, WI.

NON-CARCASS	40					
HAIR (WOOL) & SKIN	7	18				
INGESTA & EXCRETA	6	15				
BLOOD	8	20				
Plasma	4	10	51			
Erythrocytes	4	10	49			
ORGANS	4	10				
Nervous System brain, spinal cord	<	1	6			
Tongue	1	1	6			
Lungs & Trachea	1	3	31			
Heart	1	1	10			
Liver & Gallbladder	1	3	27			
Pancreas	<	<	4			
Spleen	<	<	4			
Urinary System kidney, bladder	<	1	6			
Reproductive Tract	<	<	4			
Endocrine Glands pituitary, thyroid, parathyroids, adrenals, genital (testes, ovaries), (thymus)	<	<	2			
ALIMENTARY CANAL	5	13				
Esophagus	<	1	4			
Stomach abomasum (rumen, reticulum, omasum)	3	7	54			
Small Intestine duodenum, jejunum, ileum	1	3	26			
Large Intestine cecum, colon, rectum	1	2	16			

CARCASS	60					
MUSCLES	36	60				
Neck	1	2	4			
brachiocephalicus	<	<	1	26		
remainder	1	2	3	74		
Thorax	4	7	12			
diaphragma	<	1	1	12		
pectoralis profundus	1	2	3	24		
pectoralis superf.	<	<	1	9		
serratus ventralis	2	2	4	33		
remainder	1	2	3	22		
Thoracic Limb	5	8	14			
triceps brachii	1	2	3	25		
supraspinatus	1	1	2	14		
infraspinatus	1	1	2	14		
subscapularis	<	1	1	8		
remainder	2	3	6	39		
Back	9	14	23			
longissimus	4	7	12	51		
psoas major & minor	1	1	2	9		
latissimus dorsi	1	1	2	7		
spinalis	1	1	1	6		
trapezius	<	<	1	3		
remainder	2	4	5	24		
Abdomen	4	7	11			
rectus abdominis	1	2	2	21		
transversus abdominis	1	1	2	18		
obliquus ext. abdom.	1	1	2	20		
obliquus int. abdom.	<	1	2	16		
cutaneus trunci	1	2	3	25		
Pelvic Limb	13	22	36			
semimembranosus	3	5	9	24		
biceps femoris	3	5	8	22		
quadriceps femoris	2	4	7	19		
semitendinosus	1	2	3	8		
gluteal group	2	3	4	12		
gastrocnemius	1	1	2	6		
remainder	1	2	3	9		

	% of live animal	% of carcass or non-carcass	% of major subcomponent	% of minor first subcomponent	% of minor second subcomponent	% of a combination of carcass & non-carcass parts
TENDONS & LIGAMENTS	<	1	—	—	—	—
Ligamentum nuchae	<	1	75	—	—	—
Tendo calcaneus	<	<	25	—	—	—
FAT	16	27	—	—	—	73
Subcutaneous	8	13	48	—	—	35
Intermuscular	5	9	33	—	—	24
Cavity	<	<	2	—	—	2
Intramuscular	2	3	10	—	—	7
Intraskeletal	1	2	7	—	—	5
SKELETON	8	12	—	—	—	62
Appendicular	4	6	53	—	—	33
Thoracic	2	3	22	41	—	13
scapula	<	1	5	10	24	3
humerus	1	1	9	16	39	5
radius, ulna & carp.	1	1	8	15	37	5
Pelvic	2	3	31	59	—	20
os coxae	<	1	9	17	29	6
femur	1	1	11	20	34	7
patella	<	<	<	1	2	<
tibia, fibula, tars.	1	1	11	21	35	7
Axial	4	6	47	—	—	29
Sternum	<	<	6	13	—	4
Costae	1	2	15	32	—	10
Vertebrae	3	4	26	55	—	15
cervical	1	1	7	14	25	4
thoracic	1	2	9	19	34	5
lumbar	1	1	7	16	29	5
sacral	<	<	2	5	10	1
coccygeal	<	<	1	—	2	<

	% of live animal	% of carcass or non-carcass	% of major subcomponent	% of minor first subcomponent	% of minor second subcomponent	% of a combination of carcass & non-carcass parts
FAT	6	14	—	—	—	27
Mesentery	4	10	72	—	—	19
Perirenal	1	2	14	—	—	4
Intraskeletal	1	2	14	—	—	4
SKELETON	4	10	—	—	—	38
Skull & teeth (horns)	2	6	63	—	—	24
Coccygeal Vertebrae	<	<	4	—	—	2
Metacarpals & Phalanges	1	2	16	—	—	6
Metatarsals & Phalanges	1	2	17	—	—	6

Area of a combination of carcass & non-carcass

- % of live animal
- % of carcass or non-carcass
- % of major subcomponent
- % of minor first subcomponent
- % of minor second subcomponent
- % of a combination of carcass & non-carcass parts

< = less than 0.5%

KEY:

The data represents all meat animals. Prototype was the bovine animal. Each number following a part is its percent of larger parts by weight. Example: scapula is <0.5% of live animal, 1% of carcass, 5% of carcass skeleton, 10% of appendicular skeleton, 24% of thoracic limb skeleton and 3% of total skeleton. Sum of percents within a group = 100%. Parts in parentheses may be absent due to species, sex or stage of growth.

TABLE 2.2
FRESH MUSCLE COMPOSITION
By permission of R. G. Kauffman, University of Wisconsin, Madison, WI.

NON-NITROGENOUS COMPOUNDS	79			
WATER	72	91		100
Free	63	79		87
Bound	9	12		13
LIPIDS	5	7		100
Triglycerides	5	7		98
Phospholipids	<	<		1
Free fatty acids	<	<		1
MINERALS	1	1		100
Potassium	1	1		46
Phosphorus	<	<		29
Sodium	<	<		12
Chlorine	<	<		8
Magnesium	<	<		3
Calcium	<	<		1
Iron	<	<		1
Zinc & other trace elem.	<	<		<

NITROGENOUS COMPOUNDS	21			
NON-PROTEINS	2	10		
VITAMINS (WATER SOLUBLE)	<	<	<	100
Niacin	<	<	<	48
Pantothenic acid	<	<	<	46
B_6	<	<	<	3
Riboflavin (B_2)	<	<	<	2
Thiamin*	<	<	<	1
B_{12}, biotin, folic acid	<	<	<	<
OTHERS	2	10		100
Creatine	½	3		34
Free amino acids	½	2		21
Carnosine, anserine	½	2		21
Inosine monophosphate	½	2		18

PROTEINS	19	90		
MYOFIBRILLAR**	11	54	60	100
Myosin	5	23	26	43
Actin	2	12	13	22
Titin	1	4	5	8
Tropomyosin	1	3	3	5
Troponin	1	3	3	5
Nebulin	<	2	2	3
C-Protein	<	1	1	2
M-Protein	<	1	1	2
α-actinin	<	1	1	2
Others	1	4	5	8
SARCOPLASMIC	6	26	29	100
Glyceraldehyde phosphate dehydrogenase	1	6	6	22
Aldolase	1	3	3	11
Creatine kinase	1	2	3	9
Enolase	1	2	3	9
L. dehydrogenase	<	2	2	7

CARBOHYDRATES	1	1		100
Lactic acid	1	1		75
Glucose-6-phosphate	<	<		13
Glycogen	<	<		8
Glucose, etc.	<	<		4

VITAMINS (FAT SOLUBLE) A, D, E & K	<	<		<

- % fresh muscle
  - % of nitrogenous or non-nitrogenous compounds
    - % of major subcomponents of nitrogenous compounds
      - % of each compound or element within a subcomponent

< = Considerably less than 1%

* = Thiamin is about 8 x higher in pork muscle

Other Nucleotides	<	1		6

MUSCLE AMINO ACID COMPOSITION

ESSENTIAL	MOLE %
Arginine	7
Histidine	3
Iscleucine	5
Leucine	8
Lysine	8
Methionine	3
Phenylalanine	4
Threonine	4
Tryptophan	1
Valine	6
Sub-total	49
NON-ESSENTIAL	
Alanine	7
Aspartic acid + Asparagine	9
Cystine + Cysteine	1
Glutamic acid + Glutamine	15
Glycine	7
Proline	5
Serine	4
Tyrosine	3
Hydroxproline	<
Sub-total	51

TOTAL = 100

Other glycolytic enzymes	1	4	4	15
Myoglobin	<	1	1	4
Extracellular	<	2	3	8
Others	1	4	4	15

STROMA	1	5	6	100
Collagen	1	5	6	95
Elastin	<	<	<	5
Reticulin	<	<	<	<

GRANULAR (Sarcoplasmic reticulum, mitochondria, T-tubules, plasmalemma, other membranes)	1	5	5	

MUSCLE ELEMENTAL COMPOSITION

ELEMENT	WT. %
Oxygen	73
Carbon	14
Hydrogen	10
Nitrogen	3
Sulfur & other inorganic elem.	<
TOTAL =	100

+ Data is for mature, post-rigor mammalian muscle representing various species. Each number following a part is its approximate percentage of larger parts by weight. For example: Myosin is about 5% of fresh muscle, 23% of nitrogeneous compounds in muscle, 26% of muscle proteins and 43% of myofibrillar proteins.**

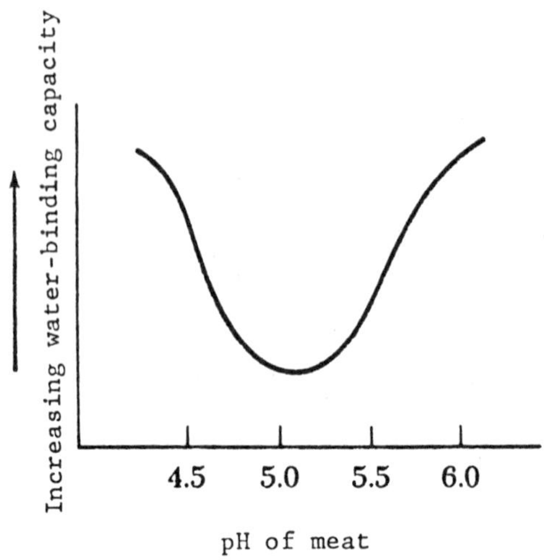

FIG. 2.1. A PLOT OF WATER-BINDING CAPACITY VERSUS pH OF MEAT
From J. Wismer-Pedersen. 1987. Part 5—Water. *In*: The Science of Meat And Meat Products, 3rd Edition. (J.F. Price and B.S. Schweigert, eds.) Food & Nutrition Press, Inc., Trumbull, CT.

Meat contains a variety of minerals including calcium, phosphorus, sodium, potassium, iron, manganese, copper and zinc. From a nutritional standpoint meat is recognized as a good source of phosphorus, iron and zinc but a relatively poor source of calcium.

The major carbohydrate of muscle is glycogen, a polymer of glucose. In the live animal there may be about 1% glycogen present, but it degrades to only trace levels during postmortem conversion of muscle to meat. The decrease in pH resulting from metabolism of glycogen to lactic acid affects not only the physical properties of the meat but also the environment for growth of microorganisms.

Meat is not a good source of fat-soluble vitamins A, D, E and K. Similarly, it is not a good source of vitamin C. However, meat is an excellent source of the B vitamins; pork is an especially good source of thiamine.

The protein component of meat is of great interest. It plays an important human nutritional role as the source of essential amino acids, is the key functional ingredient in meat processing and is the biologically functional part of muscle being responsible for contraction. In commerce, the protein composition and content determine use and price of the meat raw materials. In science the proteins are of enormous interest because of the large number present, their widely varied properties and the fact that they contribute specifically to structure and physiological function. Details about protein composition are given in Table 2.2.

Muscle proteins are divided into three broad categories—myofibrillar, sarcoplasmic and connective. Respectively, these comprise about 60, 30 and 10% of the total meat protein.

The myofibrillar proteins make up the structural part of the muscle cell (see following section on structure of muscle) and are responsible for contraction. These proteins have been very intensively studied and a great deal is known about their size, structure and properties. They are generally extractable in salt solution at an ionic strength of about 0.3, which corresponds well with the 2 to 3% salt concentration typically used in manufacture of processed meats.

Myosin is the principal myofibrillar protein, making up over 40% of the total, and it is located in the thick filaments. Actin is the myofibrillar protein of second highest concentration and is found in the thin filaments along with tropomyosin and the troponin complex. Titin is also a major component of the filament structure and alpha-actinin is a major component of the Z line. In addition, a number of other proteins including nebulin, M-protein, C-protein, desmin, F-protein, I-protein and filamin have been isolated and characterized. The specific localization and arrangement of these proteins into the filanentous structure of muscle responsible for contraction is a biological wonder.

The sarcoplasmic proteins are a widely divergent group and are extractable with water or very low ionic strength salt solutions. There are numerous glycolytic and other enzymes that fit into this category. Myoglobin, another sarcoplasmic protein, is a heme protein capable of binding oxygen and is largely responsible for the color of meat.

Finally, the connective tissue proteins represent another specialized group. They are pervasive throughout muscle and are responsible for holding the cells together, providing an extracellular skeleton and contributing to the general shape. They are insoluble, for the most part, and contribute to toughness of the muscle. Collagen is the major connective tissue protein, followed by elastin. This classification usually contains also so-called particulate proteins, which are those associated with particulate components of the cells, such as mitochondria. Collagen has an unusual amino acid composition containing more than 12% hydroxyproline. Elastin is somewhat elastic, as the name implies, and therefore found in tissue where some flexibility is required.

STRUCTURE

The structure of muscle is quite exquisite and is known in amazing detail. Individual muscles are designed to perform specific functions ranging from strong power movements, such as lifting a heavy weight, to finely graded movements, such as a manual dexterity task. Muscles have different shapes suiting them to their function, and they are often arranged in groups which act together or in opposition to other muscles or groups of muscles.

Muscle comprises about 40% of total body weight. So aside from its function of contraction it is also important in the total metabolism and shape of the body.

Muscle is one of four histologically classified tissues. The others are epithelial tissue, connective tissue and nervous tissue. Epithelial tissue occurs as a surface or lining, such as the skin or the interior lining of the digestive tract. It protects and may take on a secretory function. Connective tissue is composed of both cells and protein fibers. It binds, connects and forms a framework. It is pervasive throughout the body. Nervous tissue is specialized for memory and transmission of signals, and occurs both in the central nervous system as well as peripherally. Within the category of muscle there are three kinds—skeletal or striated, cardiac and smooth or nonstriated.

Cardiac muscle is specialized for the function of pumping blood around the body. It is composed of a syncytium of striated cells or fibers, and the most easily recognized histological characteristic is the occurrence of intercalated disks. Mitochondrial content of cardiac muscle is high and therefore it has a dark, reddish-brown color arising from the cytochromes present in the mitochondria. Cardiac muscle may be used directly as a human food, or in some cases, depending on regulatory and labeling considerations, may be incorporated into processed meats for the purpose of improving color.

Smooth muscle is composed of uninucleate, spindle-shaped cells that are contractile but which do not show the striated pattern of cardiac and skeletal muscle cells. Smooth muscle cells are very often arranged in two layers or sheets occurring around a tube or vessel, such as a component of the digestive or circulatory tract. One layer runs in a longitudinal direction and the other encircles the vessel, and this arrangement allows peristaltic movement when the two layers contract. Smooth muscle is involuntary and can be thought of as controlling or regulating the internal environment of the body. It is not normally used as a food source directly. However, smooth muscle is found in meat, for example, because skeletal muscle contains blood vessels that in turn contain smooth muscle.

Skeletal, striated or voluntary muscle is the main component of meat. It is attached to the bones or skeletal framework in order to allow transformation of contraction to movement. It is termed striated because it has a cross-striated or banded appearance, which is related to its ability to contract or shorten. And, finally it is voluntary, meaning the organism can voluntarily cause it to contract as contrasted to smooth and cardiac muscle which are involuntary.

Skeletal muscle may occur as a pennate or fusiform shape. The pennate or feather shape may be quite complicated, while the fusiform or spindle shape is much easier to visualize. The fusiform shape has two tapering ends that serve as attachments, for example, to bone, and it has longitudinally oriented fibers. Such an arrangement is obviously designed to permit a contraction to be translated to a movement.

Each muscle must also have contact with the circulatory and nervous systems. The circulatory system brings nutrients and oxygen and takes away waste products, while the nervous system transmits a signal for contraction. In the fusiform-shaped muscle both the circulatory and nervous system enter the whole muscle in the central or belly region, and the point of entry is known as the neurovascular hilum. This makes sense biologically, since these systems can enter in the central area and then disperse throughout the muscle and towards the tapering ends.

Some trapped red blood cells remain in muscle following slaughter. Because of variation in the amount of blood trapped and the fact that there is not a way to measure it easily, it is difficult to assess to what extent the hemoglobin contained in the trapped red blood cells affects color and iron content.

Illustrated in Fig. 2.2 is the fact that great detail is known about the structure of muscle ranging from what can actually be seen with the naked eye to the molecular level, which takes into account arrangement of individual protein molecules.

If a typical fusiform muscle is cut in cross section and then viewed with the naked eye the impression is that the surface actually has some structure, which is often defined as texture. The texture may range from coarse to fine. In large part this texture is a reflection of the fact that the muscle is composed of bundles (fasciculi) or clusters of muscle fibers. Two bundles are illustrated in Fig. 2.2. Bundles may vary in size and contain, perhaps, somewhere between 50 and 250 muscle fibers. The bundle is surrounded by a layer of perimysial connective tissue. In fact, there is an extensive and very well organized system of connective tissue in muscle. An entire muscle is surrounded by a layer of connective tissue termed the epimysium. Another layer of connective tissue called the perimysium segments or divides the muscle into the bundles, and yet another much finer layer called the endomysium surrounds each individual muscle fiber (also called myofiber). These layers provide a very extensive skeleton or framework, which holds the muscle together in a shape, allows transmission of contraction of the muscle fibers to anchor points so that movement can occur, and provides a convenient structure through which the circulatory and nervous systems have pathways to distribute throughout the muscle and eventually make contact with each individual muscle fiber.

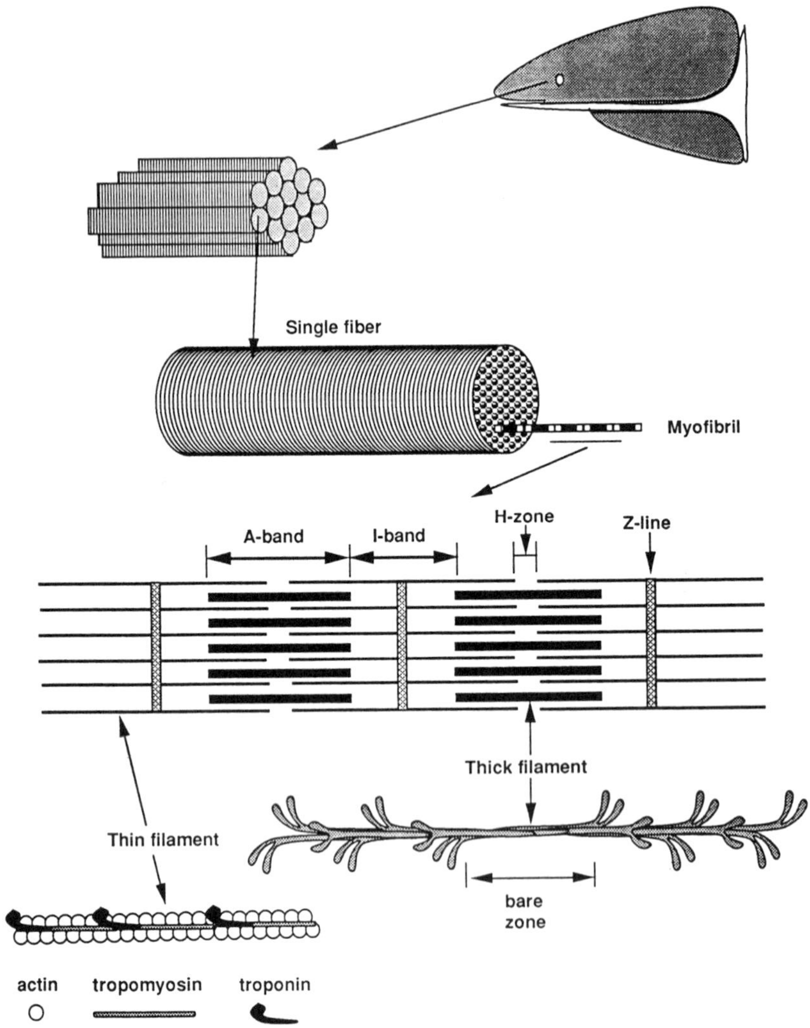

FIG. 2.2. A DIAGRAMMATIC REPRESENTATION SHOWING STRUCTURE FROM AN INTACT MUSCLE TO THE MOLECULAR LEVEL
Reproduced by permission of D.R. Swartz, University of Wisconsin, Madison, WI.

The layers of connective tissue are formed from fibers that are composed primarily of the protein collagen. Various connective tissue cells are also present. Fibroblasts synthesize collagen and export it from the cell where it then forms into the fibers. Other cell types may be present, including macrophages.

Connective tissue is also specialized into tendons and ligaments. The former attach muscle to bone and the latter connect bone to bone.

Fat cells are located in the perimysial planes and do not ordinarily occur within the muscle bundle. If a collection of fat cells becomes large enough it can be seen by the naked eye and is known as marbling.

The individual muscle fiber is a large, long cell whose distinguishing feature is the cross-striated or banded pattern. A typical mammalian muscle fiber is about 50 microns in diameter and may be up to several centimeters in length. Due to the enormous size of the cell, it is multinucleate, and the nuclei are located peripherally in the cell and just under the cell membrane. The cell membrane is termed the sarcolemma. The terms muscle cell, muscle fiber and myofiber are used interchangeably.

ULTRASTRUCTURE AND FUNCTION

As shown in Fig. 2.2, the muscle fiber is packed with smaller units called myofibrils, which are in fact the contractile units. Electron microscopy has revealed that the myofibrils are composed of sets of thick and thin filaments arranged in a regular pattern or array.

The filaments are arranged in an overlapping array, which gives rise to dark and light bands known, respectively, as A and I bands. These bands of the filaments are in register or synchrony from myofibril to myofibril and give rise to the banded appearance of striated muscle. The Z line bisects each I band. The filaments slide past each other during contraction, which shortens the distance from Z line to Z line. So a measurement of the distance from Z line to Z line (known as sarcomere length) gives a measure of state of contraction.

All of this filament structure is bathed in the sarcoplasm of the cell, which consists primarily of glycolytic enzymes and of the protein myoglobin.

Two organelles are especially evident in muscle cells—the mitochondria and the sarcoplasmic reticulum. The mitochondria produce energy by aerobic metabolism. They are located immediately beneath the sarcolemma and are also associated with the myofibrils throughout the muscle. The myofibrils are surrounded by in an extensive membrane system known as the sarcoplasmic reticulum, which functions to bind and release calcium. In association with the sarcoplasmic reticulum is another membrane system known as the T tubule and that is connected to the space exterior to the cell. When the nervous system sends a signal and depolarization of the cell membrane occurs, the signal is transmitted to the interior of the cell via the T tubule. This causes the sarcoplasmic reticulum to release calcium ions that elicit changes in the muscle proteins, resulting finally in contraction. The muscle relaxes when the sarcoplasmic reticulum binds the released calcium.

There are special muscle fibers known as intrafusal fibers that occur in clusters of only a few and are encased in a connective tissue capsule. The whole is called a muscle spindle, and the function is to act as stretch receptors.

Recent research has been directed at trying to determine the role of intracellular structures, which assist in maintaining cell shape, connect organelles to each other and attach to the cell membranes. This system is known as the cytoskeleton of the muscle fiber. There are three classes of filaments constituting the cytoskeleton, and these are known as microfilaments, intermediate filaments and microtubules. Even though knowledge is still incomplete, it is apparent that three of the proteins that participate in this interconnecting system are titin, nebulin and desmin.

The relationship between the nervous system and striated muscle is perhaps best described by definition of the motor unit. The motor unit is composed of a nerve cell body, the axon and numerous branches it makes and the number of individual muscle cells it contacts and controls. From the nerve cell body in the spinal cord an axon exits and courses peripherally to a given muscle. The axon enters the muscle and begins to branch until it makes contact with a number of muscle fibers (normally from 50 to 200). These muscle fibers are not grouped or clustered together, but instead are located individually across a wide area in the muscle. When a signal to contract originates in the nerve cell body all of the muscle fibers in the motor unit are caused to contract. Another important fact is that all of the muscle fibers in the motor unit have the same properties (such as red or white—see following section). This means that not only does the nerve transmit the signal to contract, but it also determines the properties of the muscle fiber.

Contraction is the final result of the integration of several processes. In the first place, the connective tissue functions as a skeleton for the myofibers and attaches the muscle to something so that contraction or shortening can be translated into movement. The circulatory system services the cell, which in turn produces energy. The nervous system generates a signal that is taken to individual fibers, causing a membrane depolarization. This is in turn transmitted to the interior of the cell where the sarcoplasmic reticulum releases calcium at the level of the myofibrils. Finally the filaments slide past each other, resulting in a shortening or contraction.

MUSCLE FIBER TYPES

An important circumstance in muscle is the situation that the muscle fibers are not homogeneous. In fact, the properties differ greatly, and there are two broad categories of muscle fibers classified generally as red and white. Red and white muscle fibers differ greatly in function, composition and structure. As with most biological phenomenon there is a more or less continuous range in properties between the extremes of red and white.

Figure 2.3 illustrates a histochemical test to define muscle fiber types.

The proportion of red and white muscle fibers present determines the gross appearance and properties of the muscle. In pork, for example, different muscles are clearly different in color when the cut surface of the whole ham is viewed. This difference is not so evident in beef because it is somewhat masked by the overall higher myoglobin content. The difference is clearly evident, however, in poultry where the breast muscle is white and the leg and thigh muscles are red.

Red muscle is more tonic in action than white, which is phasic. Red muscle is slower contracting, better capillarized, more aerobic in metabolism and can function more continually over a longer period of time. White muscle is just the opposite. So, as an example, a long-distance runner would depend more on red muscle to function at a given level of performance over an extended period of time, whereas a weight lifter would depend more on white muscle to do a short, very demanding task after which a rest would be required.

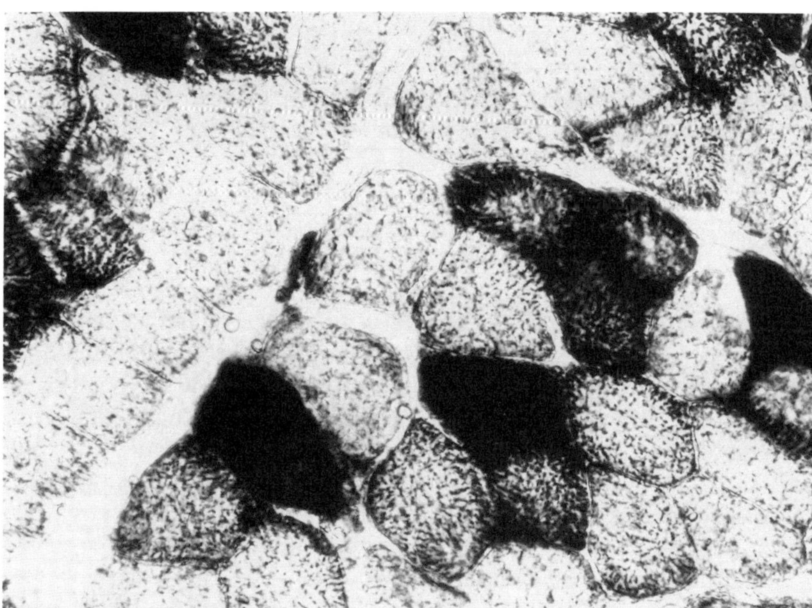

FIG. 2.3. PHOTOMICROGRAPH ILLUSTRATING FIBER TYPES IN SKELETAL MUSCLE
Red fibers are dark and white fibers are lighter. Frozen section reacted for NADH tetrazolium reductase; approximate magnification is 320x. Reproduced by permission of P.E. Mozdziak, University of Wisconsin, Madison, WI.

White muscle has lower myoglobin, higher glycogen, lower lipid and a higher concentration of glycolytic enzymes. Red muscle is smaller in diameter, has more mitochondria and a thicker Z line.

GROWTH

Embryologic development of muscle begins when myoblasts fuse to form a myotube. The myotube has centrally placed nuclei and contractile proteins are arranged peripherally. As the myotube matures, myofibrils form and take up more space, the nuclei migrate to the peripheral position and the fiber begins to take on the adult appearance. In meat animals at the time of birth, few or no myotubes (nuclei have already migrated to the peripheral location and the fiber has taken on an adult-like structure) are visible and growth occurs by enlargement of the developing fibers. During the early weeks of development of meat animals the fibers differentiate clearly into red and white types and soon reach the proportions evident in the mature individual. Growth is essentially by hypertrophy or increase in size of fibers rather than by hyperplasia or increase in number. In senescence, fibers atrophy and the total number may decline.

Satellite cells are peculiar to muscle and with ordinary light microscopy cannot be distinguished from muscle cell nuclei. They are located just peripheral of the sarcolemma but within the basement membrane. Their function is to act as a reservoir of nuclei and serve in case of damage and repair, or in growth of the fiber.

POSTMORTEM CHANGES

The postmortem conversion of muscle to meat is a complicated and critical affair. While the process itself begins with stunning and exsanguination of the animal, the treatment and handling of the animal leading up to this point cannot be overlooked. Stress imposed on the animal can indeed influence the course of postmortem events—as can handling of the carcass in the period soon after slaughter.

The immediate consequence of exsanguination of the animal is termination of the blood flow to the muscle and therefore cessation of the oxygen supply. This results in a shift of metabolism in the muscle from aerobic to anaerobic and consequently a less efficient production of adenosine triphosphate (ATP), which is the main energy source for many cellular functions, including contraction of muscle. As metabolism continues there may be an initial slight rise in temperature as the homeostatic mechanisms can no longer deal with it, but then the temperature in the muscle begins to decline.

The changes of importance are shown in Fig. 2.4. Degradation of creatine phosphate begins immediately. Acid-labile phosphorus (which is a measure of ATP) remains at a plateau concentration for a time and then also begins to decline. This more rapid fall of ATP occurs as creatine phosphate reaches a lower level and therefore ATP is no longer resynthesized by transfer of high energy phosphate from creatine phosphate to adensosine diphosphate (ADP). ATP declines as glycolysis can no longer, on its own, maintain it at a high level. pH begins to fall immediately as glycogen is metabolized to lactic acid. Extensibility begins to decline, which is an indication of the onset of rigor mortis.

Rigor mortis is the transformation of muscle from a soft and flexible state to a rigid and inextensible state. It has usually been measured by simply applying and removing, at set intervals, a weight from a strip of muscle. A recording is made of the length changes over time and the loss in extensibility as the muscle goes into rigor mortis is quite evident. The stiffening associated with rigor mortis results from the formation of links or cross bridges between actin and myosin in the absence of ATP.

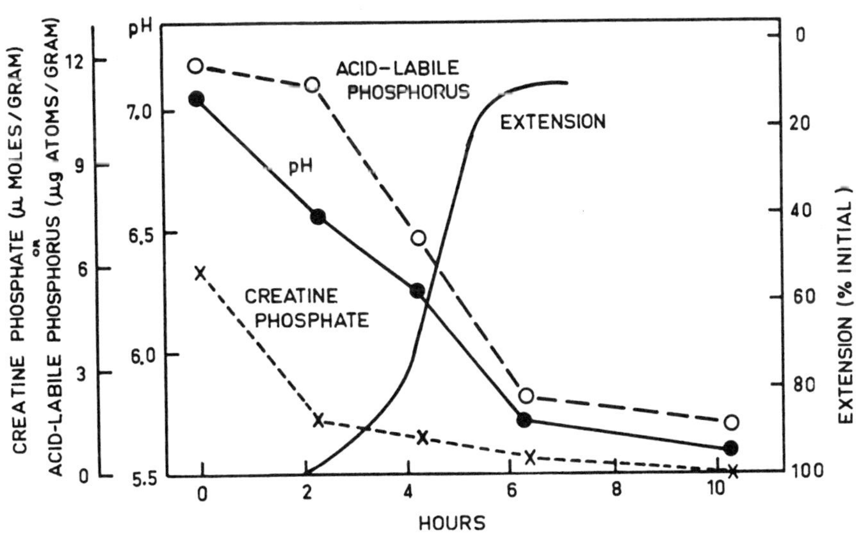

FIG. 2.4. CHEMICAL AND PHYSICAL CHANGES IN BEEF STERNOMANDIBULARIS MUSCLE HELD AT 37°C

From R.P. Newbold, 1966. Changes associated with rigor mortis. *In*: The Physiology and Biochemistry of Muscle as a Food. (E.J. Briskey, R.G. Cassens and J.C. Trautman, eds.) University of Wisconsin Press, Madison, WI. Reproduced by permission of the University of Wisconsin Press.

The changes discussed above are completed relatively quickly postmortem. Rigor mortis in beef occurs in 10 to 12 hours. In pork it takes 2 to 6 hours and has a quicker time course still in poultry. Rigor mortis is associated with a stiffening or hardening of the muscle, and so-called resolution of rigor is a softening or relaxation.

Postmortem changes have considerable variation in rate and extent, have species characteristics and may be influenced by both pre- and post-mortem conditions. The changes can play a strong role in meat quality, and this is illustrated by Fig. 2.5. The rate and extent of pH change in pork postmortem influences greatly the appearance and quality of the resulting meat. A rapid pH decline to a low level results in pale, soft and exudative (PSE) meat. Not only is it unsightly to consumers, but also it loses nutrients through drip and has poor functionality for further processing. If pH stays high the meat will be dark, firm and dry (DFD). This meat has a high water-binding capacity, but the color may not be so desirable for consumers. Microbial spoilage may occur more rapidly at higher pH.

COLOR

Color is a useful indicator of meat quality. The intensity of color is used to evaluate the age of the animal; meat from older animals contains more myoglobin, is darker and, normally, is tougher. Dull, uneven and brown color indicates bacteriological growth to the potential purchaser.

Because color is such a valuable indicator of meat quality and condition, a great deal is known about it. Although physical properties such as water-binding and textural characteristics may influence the color perceived, the major contributor to color is the heme-containing protein myoglobin. Other pigments, such as the cytochromes, and hemoglobin present in red blood cells remaining in the muscle, may make a minor contribution to meat color. It is also important to realize that meat color is essentially a surface phenomenon and that the situation only a few millimeters below the surface can be quite different.

The function of myoglobin in muscle is to store oxygen. The iron of the heme group has six coordination sites—four are used for binding to the protoporphyrin, one is attached to the apoprotein and one is available for binding. The heme iron can exist in either a reduced or oxidized form. Myoglobin is more correctly called deoxymyoglobin when there is no bound ligand at the sixth coordination site. This is the case in freshly cut meat, and the pigment in such form gives rise to a purplish-red color. If oxygen is bound, the pigment is called oxymyoglobin and the color is a bright, cherry-red, which

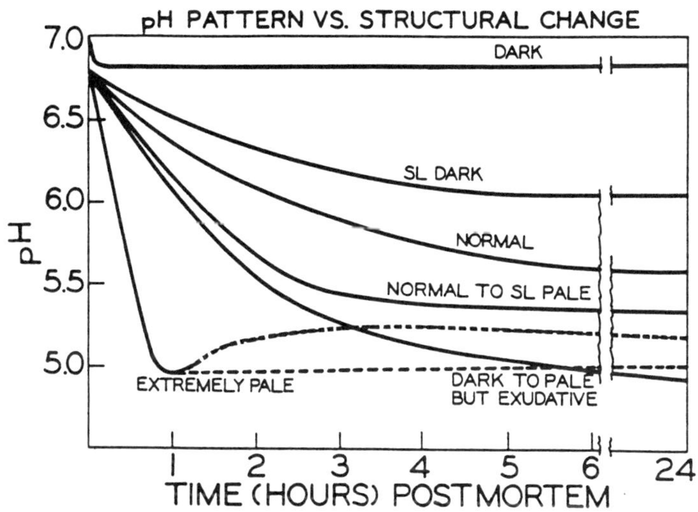

FIG. 2.5. pH PATTERNS IN PORK MUSCLE
From J. Wismer-Pedersen and E.J. Briskey, 1961. Rate of Anaerobic Glycolysis Versus Structure in Pork Muscle. Nature *189*, 318. Reproduced by permission of *Nature*.

is characteristic of fresh meat that has been allowed to "bloom". The iron is in the reduced form in both deoxymyoglobin and oxymyoglobin. If the heme iron is oxidized, the pigment form is metmyoglobin, which is an undesirable brownish-red color. Metmyoglobin is functionally inactive, incapable of binding oxygen and the sixth coordination site is occupied by water. In cured meat, nitrite is added and a series of reactions results in nitrosation of the myoglobin. When heated the pigment formed is dinitrosylhemochrome, which has a characteristic pink color. Heating uncured meat results in the grey-brown color of denatured globin hemochrome. Some alterations to the pigment, such as from certain bacteria, result in various green colors. These interrelationships are shown diagrammatically in Fig. 2.6.

A host of factors including temperature, pH, lighting conditions and extent of bacteriological growth influence the state of the pigment. Oxygen is a crucial factor, and the important relationships for fresh meat are shown in Fig. 2.7. It is clear that in order to have the desirable oxymyoglobin form and minimize metmyoglobin, the package must either totally exclude oxygen or allow it to be present at saturating concentrations.

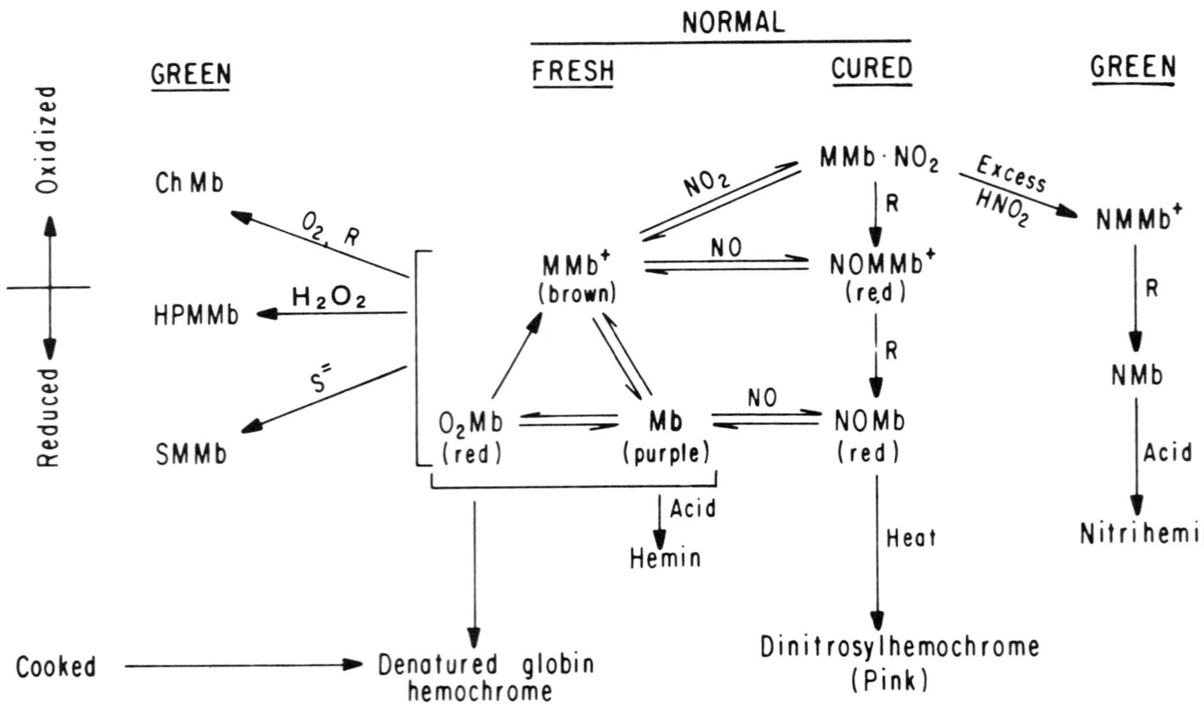

FIG. 2.6. THE HEME PIGMENTS OF MEAT

Abbreviations: ChMb, cholemyoglobin; HPMMb, hydroperoxymetmyoglobin; Mb, myoglobin; MMb^+, metmyoglobin; NMb, nitrimyoglobin; NOMb, nitrosyl (or nitric oxide) myoglobin; O_2MB, oxymyoglobin; R, reductants; and SMb, sulfmyoglobin. From J.B. Fox, Jr. 1987. The pigments of meat. *In*: The Science of Meat and Meat Products, 3rd Edition. (J.F. Price and B.S. Schweigert, eds.) Food & Nutrition Press, Inc., Trumbull, CT.

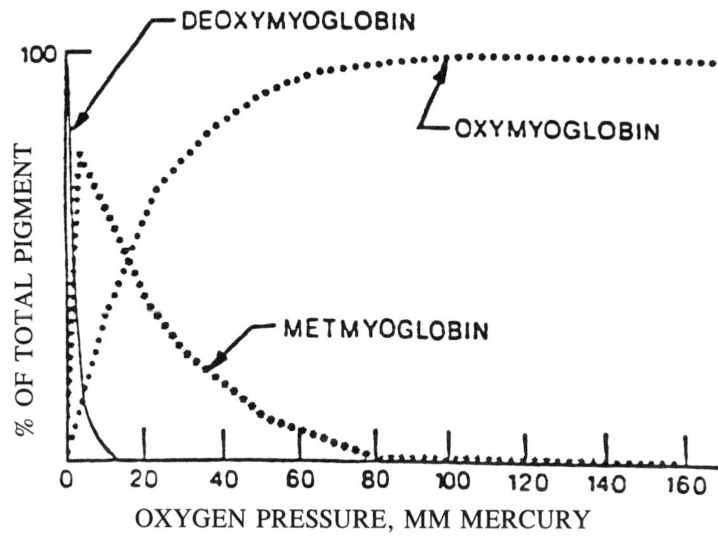

FIG. 2.7. RELATION OF OXYGEN PARTIAL PRESSURE IN THE ATMOSPHERE TO PIGMENT CHEMICAL STATES

From J.C. Forrest, E.D. Aberle, H.B. Hedrick, M.D. Judge and R.A. Merkel. 1975. Principles of Meat Science. W.H. Freeman and Co., San Francisco, CA. Reproduced by permission of W.H. Freeman and Co.

UTILIZATION OF MEAT SCIENCE INFORMATION

A basic comprehension of muscle structure, meat composition and quality characteristics is important to understanding when to use certain preservation techniques and why they work or fail.

Study of structure reveals where various components are located. It may be important, for example, that some small amount of the water present is quite tightly bound to the proteins, or that the majority of lipid present is contained within a cellular structure. Connective tissue layers may set up a barrier, or other components of structure may provide channels for invasion. The structure defines compartmentalization, which may well have an influence on effectiveness of various attempts at preservation. Structure is also related to the concept of surface area, which may be an important factor in spoilage.

Composition of meat may vary extremely depending on numerous factors. In turn, the effectiveness of various preservation procedures is influenced by composition. For example, fatty acid composition of two meats may vary considerably, and the more unsaturated will be more susceptible to oxidation and the development of off-flavors.

Quality of meat, likewise, plays an important role in effectiveness of attempts to preserve it. A low pH gives an unfavorable environment for growth of spoilage organisms while a high pH allows more rapid growth. Therefore, different quality meats respond differently to preservation attempts.

Structure, composition and quality are subject to alteration or control, thereby providing an opportunity for the individual understanding both basic science and practical operation of the industry to use preservation methods to greatest advantage.

CHAPTER 3

THE MEAT INDUSTRY

Facts about the composition, structure and function of muscle, together with a consideration of how it is converted postmortem to meat have been discussed in the previous chapter. There are a myriad of steps and procedures involved, however, from conception of the animal to arrival of the prepared meat at the table of the consumer. A broad knowledge of these steps is prerequisite to a complete and integrated understanding of the possibilities and limitations of meat preservation.

Of great importance is the fact that these production, slaughter and processing steps can influence significantly the appearance of meat and susceptibility of it to spoilage. The response to or effectiveness of application of preservation procedures may well differ depending upon how the meat was produced and processed. The purpose of this chapter is to provide a simple and general overview of animal production practices, slaughter, procedures for further processing of meat and distribution to the consumer. The rationale is to set the stage for discussion of how preservation procedures are employed during these conversions and to indicate how the more or less routine operations can indeed influence the success of attempts at preservation. As the industry achieves more refinement, and as greater effort is made to control quality and safcty of thc final product, it has become exceedingly clear that successful programs must consider the entire spectrum from the farm, through slaughter, processing and retail sale to the ultimate use by the consumer.

ANIMAL PRODUCTION PRACTICES

The process of meat production really begins with animal production practices. In fact, genetic selection pressure applied in the animal population may greatly influence composition and quality characteristics of the meat. Normally, several generations are required to make real achievements using classical quantitative genetics. Now, we are faced with the possibility of making rapid and dramatic changes using genetic engineering techniques.

Management practices, nutritional regimen, confinement and (or) animal well-being systems, and disease control programs may lead to altered meat characteristics. Such may be manifested directly as in compositional differences in the meat or indirectly through modified physiology of the animal, resulting finally in altered quality of the meat.

An example of the influence of animal production practices on meat quality is what the historical record reveals about production of PSE pork (see Chapter 2). There was a great effort by breeders of pigs during the 1950's to reduce the fat content and increase the amount of muscling in the carcass. They were quite successful. However, as subsequent generations of pigs became more muscular it was discovered that some of the best of these were producing meat of inferior quality (i.e., PSE). The pig breeders responded with genetic selection against the condition and were able to lower its occurrence. The message is that great strides can be made in selection for a desired meat characteristic, but care must be exercised that a negative impact on another characteristic is not also intensified unknowingly.

Another example is the fact that the turkey industry has had incredible success in improving growth efficiency and muscle yield through genetic selection and management practices. Recent observations have revealed, however, a rather routine occurrence of necrotic fibers in muscle of the turkey. This may be a result of the intense selection pressure for growth rate of the muscle.

Of concern in modern production practices is the use of a wide array of chemicals, such as growth promotants and stimulants and medications. There is worry that such practices may alter properties of the meat, and also there is a nagging fear about potential residues and the risk they pose to human health. Not only is there a cascading increase in the sensitivity of analytical methods, but also there is a dearth of metabolic information regarding fate of the substances used originally and conversion products resulting from them. Use of these substances is regulated, withdrawal periods are strictly enforced and routine monitoring is conducted to assure safe meat.

There are feed additives that have been demonstrated to have a very beneficial effect in the meat and are moreover of absolutely no risk to human health. For example, feeding vitamin E to beef animals results in meat that maintains a brighter color for longer periods of time and is more resistant to oxidative deterioration. So, a fresher product is offered to the consumer.

SLAUGHTER

Conversion of the live animal to meat involves a series of steps that are more or less similar for the different species. Following transportation to the slaughter facility, the animal is sacrificed, and then the hide, hair or feathers are removed. The carcass is eviscerated and chilled.

A consequential step following production of the live animal is transportation to market or the slaughter facility. Transportation can be an extremely stressful time for the animal (upsetting its physiological homeostasis) as well as possibly exposing it to disease-causing organisms from equipment or from strange animals with which they are mixed. Some European countries have devised equipment for moving and transporting animals that minimizes stress and commingling.

Similarly, practices differ in regard to the degree of comfort provided to the animals during transport and holding for slaughter. Holding prior to slaughter allows the opportunity for readjustment of physiological status, which may have been greatly altered during transportation. Another consequence is the fact that transported animals may begin shedding organisms (such as *Salmonella*) in the feces, whereas this had not been the case in their former production environment.

A key issue in preservation of meat is to keep initial contamination low. Transportation and holding of animals allows a new and different opportunity for exposure to organisms and contamination of the hair and skin as well as systemic infection of the animal, which is probably in a stressed and weakened condition.

These events leading up to slaughter also expose the animal to potential bruises or other physical damage, such as cuts and broken bones.

It is absolutely clear that the stressful time period, immediately preslaughter, can influence greatly the quality of the meat. Also, handling of the carcass immediately after slaughter has a strong influence. Both of these two crucial time periods allow an opportunity for control of meat quality, as well as the microbiological situation, and both impinge on preservation. Two examples are cited to illustrate this concept. Regarding the preslaughter period, if the animal is stressed the pH of the resulting meat may be lower or higher than the normal range and may have become an environment less or more desirable for microbial growth. Regarding the immediate postslaughter period, proper chilling is quite effective in controlling microbial growth.

The initial steps in the slaughter process are stunning and bleeding. Prior to this, however, the animal should be cleaned to the extent possible, since dirt and manure on the animal is a major source for potential bacterial contamination. The animal is restrained, usually by some type of squeeze chute arrangement, and stunned by mechanical (penetration or concussion), chemical (carbon dioxide) or electrical means. The objective is to render the animal unconscious but not to kill it. Bleeding is accomplished by inserting a knife into the throat area and severing the carotid artery. The animal is normally suspended or at least on its side and the result is a sudden rush of blood from the body, which gives a rapid death. As much blood should be removed from the muscle as possible. Even under the best circumstances, however, some residual blood remains in the muscle in the form of trapped red blood cells in the circulatory system.

Unless the living animal is suffering from a disease or systemic infection, the muscle should be free from microorganisms. Since the function of the lymph nodes is to filter out and collect microorganisms, they represent, even in healthy animals, a potential source of contamination during subsequent fabrication of the carcass. Microorganisms can be introduced with the knife during the sticking process, and a chance exists that they may then be distributed via the circulatory system to the muscle. Obviously, the number would be small and the interior of the muscle would present an essentially anaerobic environment.

From this point onward then, animals generally move on a continuous rail system that passes by various worker stations where certain tasks are performed.

The next step is to remove the hide, hair or feathers. These operations represent a potential major source of contamination. Control depends primarily on ensuring that the surface of the animal is as clean as possible and on insisting that workers use procedures correctly and prevent what is being removed from the exterior from contacting the surface of the carcass as it is freshly exposed. Since the workers are handling the hide as they skin the animal it is difficult to avoid transfer of contamination as they touch the freshly skinned areas. Mechanical devices for pulling the hide are being used more commonly and minimize contamination. In the case of pigs and poultry, the usual procedure has been to remove the hair or feathers and leave the skin on the carcass. This is done by scalding the carcass in hot (145°F) water and then scraping, usually with a mechanical device. To ensure complete removal of hair from pig carcasses, hand scraping, resin depilation or singeing may be employed.

Evisceration is another potential source of microbiological contamination. The digestive tract is loaded with bacteria and any puncture or spillage can result in a heavy microbiological contamination of the carcass, making subsequent preservation more difficult.

The carcass is normally split (or opened in poultry) and then cleaned by washing. Weighing and some grading or evaluation of the carcass may occur prior to chilling, or subsequently. Large variations in chilling efficiency are found among chilling rooms in which carcasses are normally suspended. The type of system and loading of the room are major influences. With poultry, chilling is normally accomplished by immersion of the carcass in a chilling tank or bath. Strong interest in obtaining the best chilling conditions possible is becoming more evident, and cryogenic possibilities are being explored.

From the standpoint of subsequent preservation of the meat, the two most important factors are initial level of contamination of the carcass during the slaughter and dressing procedures and temperature control. The former depends upon the level of hygiene practiced by the plant and the latter depends primarily on type and operation of the chilling system. The desire to reach a low temperature rapidly must be balanced against possible negative effects on yield and palatability. It has been established, for example, that chilling too rapidly can cause cold-shortening of the muscle and associated toughening.

Because of the recognized importance of keeping initial contamination as low as possible, there has been considerable interest in using organic acids (such as acetic, citric and lactic) to rinse or spray the carcasses as a means to inhibit growth of both spoilage and pathogenic organisms (see Chapter 6 for details).

After the carcass has reached a sufficiently low temperature it is fabricated into smaller units. This is normally performed the day following slaughter but may be sooner in some plants employing rapid chilling. In some circumstances so-called hot-boning may be practiced. Muscle groups are removed soon after slaughter (prerigor) and then used immediately, for example, to produce a processed meat such as fresh pork sausage.

In the beef industry, carcasses are cut into primal and subprimal cuts and packaged in vacuum bags. These are distributed to the retailer where they are processed to retail cuts. Trimmings are distributed in the fresh or frozen state. In the case of pork, a relatively large portion of the carcass (i.e., ham, belly and shoulder) are destined for processing. The loins are generally shipped fresh (in boxes) and then cut at the retail store. Some success has been obtained in cutting at the packing plant for direct shipment to retail outlets.

The use of modern design and building materials has permitted improvements in sanitation, which is related to preservation of meat. One-story buildings have allowed for an orderly and efficient flow that can be controlled to minimize recontamination and cross-contamination among products at various stages of the process as well as among workers from different areas. Stainless steel equipment and impervious floor and wall coverings have made cleanup easier. Considerable effort has been directed at improving ergonomic design for worker comfort and safety, as well as providing proper lighting in working areas.

FURTHER PROCESSING OF MEAT

An alternative to selling meat in the fresh state is to convert it to a meat product. Further processing is somewhat difficult to define because of the extremely wide variety of products that may be made. Merely grinding meat is indeed a transformation, and the hamburger industry is obviously an important one. Similarly, more and more meat is being utilized in frozen entrees, which can also be viewed as processed. However, processing is generally considered as involving either a curing type of reaction, or heating, or some combination of both. A complete discussion of curing is given in Chapter 6 and heat processing is described in Chapter 5.

In any event, a processed meat product has quite different preservation properties compared to fresh meat. This is an important fact, and will be a recurring theme for discussion and analysis.

LEEDS UNIVERSITY LIBRARY

Meat was processed in the past with the primary intent of actually imparting better keeping qualities to it. While the preservative properties are still important, more emphasis is placed now on adding some unique characteristic or value to the meat. Processing allows the development of new or additional products that may appeal to different market segments. Or, the processing may be directed at filling a need such as convenience, demand for a low-fat product or special keeping properties.

Meat processing is an important segment of the total meat industry. Nearly two-thirds of the total pork produced is processed; 10 to 15% of beef is processed; and for poultry the amount is about one-fourth. These figures are only estimates, and processing is defined as more than merely grinding.

It may be simplest, for the purposes of discussion, to consider three broad classifications of processed product—these are cured, canned and sausages.

Ham and bacon are pork products that are cured, heated and smoked in the intact form. Salt and nitrite, which are the major ingredients in curing, impart special keeping properties, and the process of heating likewise is a form of preservation. The curing solution is normally injected, but in some special cases the dry ingredients may be rubbed on the surface and then allowed to penetrate. Such dry curing requires longer time periods. These products are generally smoked during the heating cycle. Many hams are now made from smaller pieces of meat that have been formed back together. The cook-in-bag procedure has gained increasing importance in this regard.

Canning may take the form of a pasteurization or sterilization. In the latter, the product is shelf-stable at room temperature for several years. In the former a curing step is usually included, but the canned product must still be refrigerated during storage. Canned hams comprise the larger portion of this category. There is a great variety of canned meat and meat-containing products such as stew, chili and spaghetti. They are noted for a reasonable price and excellent keeping quality.

The third group is sausage type products, which usually undergo curing or heating during manufacture. Some dried and fermented sausages are not heat processed. Ordinarily, such sausage-type products are either ground coarsely or chopped finely to form an emulsion. Curing ingredients are incorporated during the size reduction phase, the mixture is then stuffed into a casing or mold and heat processed. A smoking phase is often included. Fermented and dried sausages undergo a longer manufacturing schedule and may or may not be heat processed. Fresh sausage is only ground and seasoned.

REGULATORY CONSIDERATIONS

Governmental regulatory processes are employed at numerous points to assure safety of the meat produced, and residue monitoring programs are also conducted. The Meat Inspection Act was passed in 1906. Since then, several other legislative acts have been passed including the Poultry Products Inspection Act of 1957 (amended to Wholesome Poultry Products Act in 1968), the Humane Slaughter Act of 1958, and the Wholesome Meat Act of 1967. The latter act ensured that state meat inspection was at least equal to federal. Meat inspection is currently the responsibility of the Food Safety and Inspection Service. Basically, there are seven areas which are: ante- and postmortem inspection, reinspection (processing), sanitation, potable water, sewage and waste disposal control, pest control, and condemned and inedible material control.

It is apparent at the time of this writing that changes will be made in the federal meat and poultry inspection system. While plans have been underway to modernize and improve the system, effort was obviously intensified following a food poisoning outbreak in mid-January 1993 during which several hundred people became seriously ill and at least three deaths occurred as a direct result. The cause was hamburger that had been contaminated with *Escherichia coli* 0157:H7 and served undercooked.

The present system works well for what it was designed to do more than 80 years ago—that is, to prevent entry of diseased and damaged meat into the human consumption chain and to ensure that the operations converting food animals to meat are conducted in a sanitary and appropriate manner.

The present system depends on onsite, direct, visual inspection. The system will probably be altered to be more of a preventive type rather than after-the-fact inspection. Undoubtedly, an HACCP system (see complete description in Chapter 8) will be put in place and will encompass the total picture from farm production to consumer's table. HACCP is known to be effective in a multistep and multisite production scheme such as just described in this chapter for meat. In addition the inspections system will be more science-based, and it will employ technologies such as irradiation pasteurization, carcasses rinses and animal identification and trace-back procedures.

Undoubtedly, education of consumers will also be a major consideration in the new plan. For example, a rule proposal has already been made that packages of raw meat must include explanations of how to handle and cook the food properly.

DISTRIBUTION

Today, very little meat is distributed in carcass form (with the exception of poultry). Rather, it is reduced to primal or subprimal cuts and then distributed to further processors, purveyors or retailers. Once the fresh meat is reduced to retail package size the shelf-life is limited to only a very few days. On the other hand, processed meats ordinarily leave the manufacturer in a package ready for retail display. Distribution chains may be long and complicated, and shelf-life is measured in weeks instead of days. A package of wieners will have a shelf-life of six weeks or longer, for example.

At retail, the processed meats are often displayed at eye level in a vertical refrigeration unit, or in a deli-type setting. Fresh meats are normally offered from an open top or case type of refrigeration unit. In both instances the meat is subjected to lighting conditions that will display it to best advantage. A problem is the fact that temperature may fluctuate depending upon configuration of the display unit, loading and time of day.

A growing segment of the industry is devoted to food service. In this case the distribution is ordinarily direct from the manufacturer or via a purveyor. There are normally rather strict specifications for condition (i.e., temperature) of the product upon receipt and how it is handled and prepared.

When the consumer takes meat from the retail store it may be stored and prepared with excellent technique or it may be abused. No one except the consumer has control once the retail purchase is made. However, the manufacturers name remains attached, as does a responsibility.

The raising of animals, converting them to meat and processing the meat to numerous products can be accomplished by an enormous number of different procedures and techniques. Distribution, and subsequent handling and preparation by the consumer further complicates the situation. All of the steps have potential of influencing preservation properties—some are well-known and others are only suspected.

CHAPTER 4

PRESERVATION AGAINST WHAT

What happens once the muscle is meat and the clock starts to run? It was explained in Chapter 3 that previous treatment of the animal is extremely important. The live animal production aspect results in meat of widely varying composition and quality properties. The slaughter of the animal not only influences quality properties of the meat, but also yields an opportunity to control hygiene and regulate the microbial contamination. Finally, the post-mortem conversion again influences quality aspects such as pH.

Once meat is meat, there are basically two ongoing processes that impinge on attempts at preservation. There are chemical and physical changes occurring within the meat itself. There is also another whole set of reactions occurring because of the dynamic microbiological population—both spoilage and pathogenic organisms may be present. Finally, there is the possible interaction of the two processes. Both the chemical-physical and the microbiological changes are situated within the given structure and composition of the meat as well as the environment to which it is exposed. Each process may influence the other.

A clear distinction must be made between what is happening in the interior of the meat and at the surface. While nothing is ever absolute in biology, it is obvious that most microbiological activity is at the surface; chemical and physical changes are in the interior as well as at the surface. While the surface is most likely muscle (often cut cross-section to the long dimension of the muscle fiber), it may also be fat or exposed layers of connective tissue. A ground or rolled form of meat presents a somewhat different situation. In the case of ground meat, there is an enormous increase in surface area as the large pieces are reduced to smaller entities, and a mixing action also occurs during grinding that brings surface contamination to the interior. In a rolled meat, part of the former exterior surface may well be incorporated into the interior.

Competition among microbial species is critical. Microorganisms compete with each other for nutrients and oxygen and also alter their immediate environment to make it unattractive to other species (see Chapter 7 for full explanation). Some compete more successfully, and therefore, the initial level and conditions for their growth are important to the final outcome. For example, the presence and growth of spoilage organisms may minimize the growth of pathogens.

We have a complicated, mixed and vigorous system in and on the meat that can be influenced by numerous factors under our control. Here then is the mission—how to understand the circumstances and control of the events about to take place. In other words, how can the meat be preserved and safety assured?

APPEARANCE AND PALATABILITY CHANGES

The biochemical basis for color has been explained in Chapter 2. Fresh meat has a bright red color and as it begins to age, it turns darker and, finally, brown. This happens as the result of chemical changes, physical drying and microbial metabolism. It is difficult to separate the ongoing color changes, and their resultant impact, from the changing microbiological situation. Even so, the appearance of brown color is often taken as an indication of microbiological spoilage. While this is a normally accepted explanation, it is not always the case, as brown metmyoglobin may result from conditions of oxygen exposure. Or a color reversion may occur in which the brown color is converted again to a red in the presence of high populations of some microorganisms.

Both species and animal age influence meat color. Muscle from older animals has more myoglobin and therefore a meat with darker color. In increasing order of myoglobin content are poultry, pork, lamb and beef. Individual muscles also vary in myoglobin content, with active muscles having more. Depending on the packaging or storage environment, desiccation may occur at the surface of meat as it ages, resulting in a darker color. Also, the physical binding of water or the exudation of moisture at the surface each influences color, giving for instance an impression of a lighter color if more loose or unbound water is present.

The classical method has been to ripen or age beef before considering it ready for consumption. Essentially, this meant to hold it (usually in carcass form) at refrigerated temperature for a period of time—about 10 days to two weeks. The desired result was a decrease in toughness and an improvement in flavor. In general, this happened and was probably the result of proteolysis.

A more aggressive form of ripening is dry aging. In this case, the unwrapped wholesale cuts (usually ribs or lions) are held for a considerably longer period of time (up to several weeks) at a cooler temperature. The result is development of stronger, more intense flavor, a substantial weight loss due to desiccation, required surface trimming and an accordingly higher price. It is a practice now nearly extinct.

The foregoing generally applies only to beef. An attempt is made to move pork rapidly in the distribution chain, since the fat is more unsaturated, more labile to oxidation and therefore more prone to development of off-flavor. Similarly, poultry is not aged, but marketed quickly.

During the last 25 to 30 years, there has been an almost complete shift away from aging beef carcasses. Now most carcasses are fabricated a day or so after slaughter and the resulting primal and subprimal cuts are packaged in vacuum bags. These are held and distributed in this form and are generally not opened until receipt at retail outlets. This distribution requires two to three weeks. In essence, the conditioning is accomplished in the vacuum bags instead of the total carcass. The surface reactions and the microbiological population of vacuum packaged primal are quite different from what would occur in fresh meat exposed to air, due to the anaerobic atmosphere in the vacuum packages.

Attempts are being made to convert carcasses directly to retail-ready packages. Extraordinary hygiene is being used, and some progress in extending self-life is being made. However, it remains difficult to obtain sufficiently long shelf-life for many distributions chains.

PROTEOLYSIS

It has long been an accepted practice that the holding of meat postmortem at chiller temperature improves palatability—most notably tenderness of beef muscle. This has been referred at as aging, conditioning or ripening. In essence, it seems that the muscle "softens". It is difficult to isolate this type of physical change from change in flavor, for example, which is going on simultaneously. Proteolysis defined here is also distinct from microbiological events.

The concept of proteolysis-induced tenderization appears quite simple. However, it has proven to be a most complex process to characterize and simulate. This phenomenon serves as an excellent historical example of how difficult it can be for scientists to unravel and explain physiological, morphological and biochemical events so that some control of the natural process may be accomplished at the commercial level.

During the first third of the 20th century, scientific investigation demonstrated that conditioning of muscle resulted in an accumulation of proteolytic products. This was substantiated in so-called "sterile" systems that eliminated the possibility of proteolytic activity due to microorganisms.

As more careful investigation was made, however, it became clear that the actual extent of proteolysis was not great. Moreover, histological investigations did not reveal the expected massive structural damage. The viewpoint then was adopted that a specific proteolysis was occurring at certain key points. This could give a weakening in the complex structure and in turn, an improvement in tenderness.

It has been observed that the Z disk degrades, which results in the myofibrils fragmenting more easily. Electrophoretic studies have shown little or no degradation of myosin, actin or alpha-actinin, while troponin T, desmin and titin are lost from detection at a gradual rate during conditioning. It is important in interpreting these studies to know if they are conducted using isolated or disrupted systems or in normally handled and intact meat. The temperature of incubation is also quite important.

There are numerous proteolytic systems in muscle, but only two appear to be important in meat. The most important system is the Ca^{2+}-dependent proteinases, also called the calpains. Of secondary importance probably are the cathepsins that are located in lysosomes.

While calpains have been demonstrated to degrade the Z disk, myosin, actin and alpha-actinin at high temperatures, they are not effective at 2 to 4°C. Calpastatin is a specific inhibitor of calpain, and may play an important regulatory role.

The connective tissue component of muscle also contributes to tenderness. There is not a histologically detectable proteolytic degradation of the connective tissue during conditioning of the meat, but a possible subtle damage to the regular and complex structure. Cathepsins may produce some depolymerization of the collagen aggregates so that they are more heat labile, and therefore the denatured fibers have less strength.

It is clear that proteolytic changes take place during conditioning. They are not, however, extensive when assessed by chemical or histological evaluation, but apparently specific, effective and delicately controlled. Factors such as pH, ionic strength and temperature may greatly influence proteolysis. These parameters are influenced by live animal production practices and slaughter procedures.

OXIDATION

Oxidation of fat results in characteristic odors or flavors denoted as stale or rancid. It is a significant problem with fat-containing foods. A complex mixture of reaction intermediates or compounds is produced and the presence of these may be fleeting, making quantification and study difficult. The end products of oxidation are, however, readily apparent to the consumer.

The extent to which oxidation occurs in meat depends on two factors—its composition and the conditions to which it is exposed. Meat with a higher unsaturated lipid concentration is more labile to oxidation. Higher temperature and increased light intensity act as prooxidants.

Technically, the process is an autooxidation. When oxygen attacks a double bond, a hydroperoxide is formed, Then it goes on as a free radical chain mechanism. This gives rise to numerous products including short chain aldehydes. These short chain aldehydes, with acids formed by their oxidation are for the most part responsible for the characteristic and undesirable flavors and odors.

Attack of lipids by lipolytic enzymes is also possible, but the fatty acids released normally do not give rise to flavor problems as do the oxidation products described above. The liberated free fatty acids are, however, more susceptible to oxidation than are the bound fatty acids.

Oxidation of lipid in meat products is especially interesting because of the close interrelationship with oxidation of the meat pigment and associated change in color. Both heme pigment and non-heme iron have been shown to influence oxidation of lipid.

Numerous antioxidants exist that slow the development of oxidative rancidity. However, two problems exist with their use—getting them to the site of oxidative reactions and complying with numerous regulations for use of such additives.

FUNCTIONALITY

Not much is known about change in functionality properties of meat as it ages. Certainly, it makes sense that the longer meat is aged, the greater the possibility for development of flavor problems. However, the question here is whether aging influences the performance of the meat if it is further processed. Does it have better or poor water-binding properties? Will fat separation occur more easily if it is made into an emulsion-type product? Will cured color be affected adversely?

Whilc it is generally accepted that there is some loss of functionality during the process of freezing and thawing meat, frozen meat is still used for manufacture of processed meat products. Common sense and prevailing opinion indicate that functional properties are less desirable in less fresh meat. In reality, there is little scientific evidence to cite in support of this idea.

SPOILAGE

Spoilage of meat is actually a natural decay that occurs due to processes in the meat itself or to contamination by microorganisms and their subsequent growth. Our main interest here is how to control or regulate the decay. Spoilage of meat means different things to different people. For example, some individuals may be quite pleased with the stronger flavor of dry aged steak while others may object or even reject it. The odor from a newly opened vacuum bag of fresh meat may be quite unpleasant, but subsequently as the odor dissipates, it is apparent that the meat has not undergone any significant spoilage that would affect its perception by the consumer.

Change in color, which is the development of brown pigments, is usually taken as the first sign that spoilage is underway. The appearance of brown metmyoglobin may not be due solely to microorganisms, although it is generally taken as an indication of condition of the meat or growth and increasing numbers of microorganisms. A surface greening may occur if microorganisms produce hydrogen peroxide (or hydrogen sulfide) that may in turn oxidize the myoglobin to porphyrins that have a green color. Off-odor is a clear warning that deterioration is well underway. On a given piece of meat, both sour and putrid smells do not generally occur at the same time. Sour implies spoilage by lactic acid organisms. A putrid odor results from aerobic storage and is generally indicative of temperature abuse. A slime may build up on the surface as visible colonies are formed; unusual colors may also appear.

The microorganisms associated with meat spoilage act primarily on the proteins present, usually after readily available energy sources such as glucose are depleted. Proteolysis converts the proteins to soluble peptides and amino acids. Subsequent microbial action on the amino acids can be deamination or decarboxylation, giving rise to such compounds as ammonia, keto acids, carbon dioxide, and various amines. Metabolism of specific amino acids can result in substances like hydrogen sulfide and indole.

There are microbial lipases that may hydrolyze fat to glycerol and fatty acids. Also, there are microbial oxidases that may oxidize fatty acids. However, autooxidation is much more important in meat than is development of rancidity due to microbial action.

Actual monetary loss due to spoilage of meat is substantial. It is estimated that spoilage accounts for 1 to 5% of the value of retail meat. This results from mark downs in price as the retailers attempt to salvage some value, and from total loss if spoilage reaches the point where the meat can no longer be used.

Meat is an extremely perishable food. Its moisture content is high and relatively available, and it has an excellent nutritional profile for microorganisms as well as humans. How rapidly it spoils depends primarily on: (1) type and load of microorganisms with which it is contaminated, and (2) the environment in which the meat is subsequently exposed.

A variety of bacteria, molds and yeasts may be deposited on the surface of the meat. In general, bacteria grow faster than molds and yeasts, and spoilage of fresh meat is therefore attributed to bacteria. The initial number of microbes deposited, the environment of the meat and the ability of a specific organism to compete determines the outcome. Affinity for substrates is also important. Another complication is the fact that the microorganisms may attach firmly to the meat surface and make removal by washing difficult.

In terms of preservation, the total amount of water in the food is not as important as the amount of available water (a measurement of what solutes are dissolved in the water). This measurement is called water activity (a_w). It is defined as mole fraction of the solvent divided by the moles of solute plus moles of solvent. Fresh meat generally has an a_w of 0.99 or above that is essentially optimum for growth of most microorganisms. An a_w of 0.85 will generally inhibit bacterial growth, while some molds may still grow at an a_w as low as 0.75.

Oxidation-reduction potential affects growth of microorganisms. Aerobic organisms require oxidized conditions, whereas anaerobic organisms need reduced conditions. The oxidation-reduction potential in muscle is positive at the time of slaughter and it declines and becomes negative during postmortem time. The interior of large chunks of meat are generally reduced, even under aerobic storage.

Almost all meat is refrigerated. Therefore problems are primarily related to microorganisms that can survive and grow at lower temperatures. However, temperature abuse must also be considered. Temperature abuse is a commonly used term meaning that the meat is exposed to ambient or elevated temperature rather than refrigeration temperature. While microorganisms have different optimum growth temperatures, increasing the temperature (within reason) nearly always increases growth rates. A usual optimum temperature classification is 55 to 80°C for thermophiles, 30 to 40°C for mesophiles, 25 to 30°C for psychrotrophs and 10 to 15°C for psychrophiles. Obviously, the organisms can exist and grow in a range of temperatures to either side of the optimum. For example, the ability of psychrotrophs to grow at or below 7°C is a far more important issue for preservation of meat than is their optimum growth temperature.

Fresh meat at retail generally has a high concentration of oxygen at the surface due to the high oxygen transmission rates of packaging materials designed to maintain myoglobin in the oxy-form. Sometimes, if the film has lower rates of oxygen transmission, the atmosphere inside the package may actually become low in oxygen. Cured meats are vacuum packaged, resulting in a nearly anaerobic environment.

Contamination of fresh meat typically originates at slaughter where the major sources of microorganisms are the hide and hooves of the animals. Contamination also occurs via the workers and the equipment they use. Contents of the digestive tract represent an enormous reservoir of microorganisms, but great care is taken to avoid accidents and spillage during slaughter. Further contamination occurs at the time of preparation for retail sale when the meat is reduced to smaller pieces. Contamination of cured meat is usually a recontamination from the environment, equipment and workers and follows heat processing. Finally, contamination may occur in the home as consumers handle the meat.

Growth of Microorganisms

A typical growth curve for microorganisms is shown in Fig. 4.1. Bacterial reproduction is by fission, which means that a cell divides and becomes two cells. Generation time is the time required for the population to double. In meat under ideal conditions, it is possible (but not usual) to have a bacterial generation time of only 20 to 30 minutes, so enormous numbers of microorganisms can arise in a relatively short time. Although it seems that the microbial population will continue to increase to infinity, it should be recognized that microbial growth will stop once nutrients are depleted and toxic waste products and metabolites build up to certain levels.

In simplest terms, there is at the outset a lag phase in which there is no change in microbial numbers. During this time, the microorganisms are preparing to divide. This is followed by a very rapid increase (exponential) in numbers of microorganisms. Then there is a plateau or stationary phase followed by a decline and die off of the population. Each organism, depending on its generation time, nutrients available, environment and ability to compete with the potentially numerous other organisms present, follows its own particular growth curve.

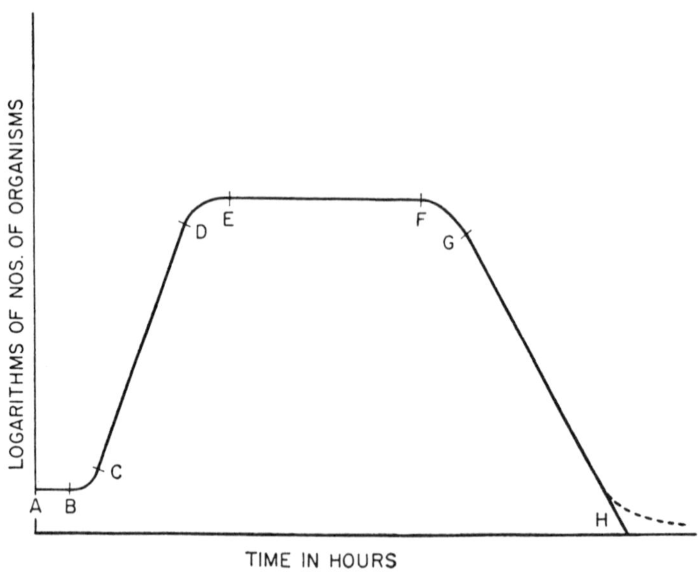

FIG. 4.1. GROWTH CURVE OF MICROORGANISMS
A to B, lag phase; B to C, phase of positive acceleration; C to D, logarithmic or exponential phase; D to E, phase of negative acceleration; E to F, maximal stationary phase; F to G, accelerated death phase; and G to H, death phase. From W.C. Frazier. 1967. Food Microbiology, 2nd Ed., McGraw-Hill Book Co., New York. Reproduced by permission of McGraw-Hill Book Co.

Most preservation methods are aimed at inhibiting growth of microorganisms rather than killing them. Optimum results are obtained if action is directed at the lag phase. In general, preservation methods extend the lag phase or alter the slope of the log phase.

Molds

Molds are recognized when growing on the surface of meat by their fuzzy appearance. Molds found on meat may range in color from white to gray to green. They are multicellular, filamentous fungi with a complex microstructure that is often used for their identification. They have a branched structure with intertwined filaments called hyphae. The hyphae grow into the food, extend up into the space above and are responsible for visible coloration. Often the color of hyphae darkens when they are bearing spores.

Reproduction is by spores. This facilitates contamination, since the spores are hearty and are very easily spread via the air.

Molds require oxygen and can grow in more acid pH environments than can bacteria. Most are mesophilic and some can grow well at refrigerated temperatures. They generally require less available moisture than yeasts and bacteria. In a competitive environment, they generally do not grow as well as other microorganisms.

Yeasts

Yeasts are unicellular or single-cell organisms, usually with a spherical or slightly elongated shape. They reproduce by budding.

As contrasted to molds, yeasts require more plentiful moisture for optimum growth. Their temperature needs are similar to molds and they also tolerate an acid pH.

Some yeasts are salt tolerant and nearly all favor sugar as an energy source. In general, they require oxygen for growth, but some fermentative types grow anaerobically.

In meat products, yeasts are often recognized as producers of surface slime.

Bacteria

There are many bacteria distributed ubiquitously in our environment. They are amazingly diverse with some being adapted to extreme conditions like hot springs. These prokaryotic cells appear as a sphere, rod or spiral and reproduce by fission. Some have capsules, some are spore forming and some have flagella providing motility. Differential staining, batteries of biochemical tests and immunological characteristics are used for classification and identification. The study of taxonomy or classification is an enormous area in itself, and a microbiology textbook should be consulted for details.

The bacteria presented on fresh meat are most likely aerobic organisms like *Pseudomonas*, *Acinetobacter* and *Morazella* species with the *Pseudomonas* generally predominating. Vacuum packaging of fresh meat restricts the supply of oxygen giving the lactic acid bacteria an opportunity to grow. In the case of cured meat, an entirely different flora comes into play. Most commonly, the following occur: *Staphylococcus*, *Micrococcus*, *Lactobacillus*, *Microbacterium*, *Pediococcus*, *Streptococcus*, *Clostridium* and *Bacillus* species.

MEATBORNE DISEASES

Besides causing spoilage of meat, microorganisms may also be responsible for illness in humans. The mode of action is by one of two routes. An infection occurs when the organism is ingested and then has the opportunity to grow within the host. An intoxication occurs when a preformed toxin is ingested. The illness resulting may range from the common upset stomach to a life-threatening condition. Diarrhea, vomiting and fever are common symptoms.

In general, foodborne illnesses have been difficult to document, and most are not reported officially. Unless the illness is serious or lasts for a period of time, most people do not consult a doctor. If a foodborne illness is suspected, the food thought to have caused it must be recovered and analyzed, and the infected people must also be tested to identify the organism responsible for their illness. This process is expensive and usually takes longer than the 24 to 72 hours necessary for the sick individual to recover. The outbreaks most frequently documented are those affecting many people and can be traced fairly easily to a specific meal or food source. Infants, the immunocompromised and the elderly are usually more susceptible to these illnesses.

Meat is a potential food poisoning vehicle. Animals are often carriers of problem organisms, either in the intestinal contents or the soil contaminating the hide. That is why proper sanitation and hygiene during slaughter and dressing procedures are so important.

Most foodborne illnesses occur because of mishandling in the home or in food service establishments. These problems often take the form of cross contamination from raw to processed or prepared foods, or from improper temperature control—either insufficient heating or refrigeration.

Due to an increase public awareness about foodborne illnesses, improved scientific understanding, rapid and accurate detection methods and governmental regulatory activity, there is now much greater documentation about the subject. Several "newly emerging" pathogens have been implicated in foodborne disease in the past decade. All facets of the industry devote considerable attention to preventing the possibility of a foodborne illness arising from their product.

There are two factors important to note, however. First, there are technologies, such as irradiation, that would improve safety of meat, but are not being used because of potential consumer resistance. Second, it is virtually impossible to guarantee 100% safety, despite precautions taken and procedures employed.

Following is a brief discussion of foodborne illnesses that may be associated with meat. Standard texts should be consulted for complete details.

Perhaps the best known foodborne illness is botulism, a potentially fatal condition. In recent years, the mortality from this disease has declined due to improved treatment procedures and the availability of an antitoxin. Cases are now relatively infrequent. The *Clostridium botulinum* organism is widespread in soil and produces a heat-resistant spore that upon germination produces the toxin. There are seven types of the organism based on antigenic specificity of the toxin. For example, type E is associated with fish while type C is usually associated with waterfowl. The toxin attacks the myoneural junction, and the symptoms of the disease are nausea, dizziness, headache, double vision, paralysis of muscle and difficulty in breathing.

The word "botulism" came from the Latin for sausage (botulus) indicating the disease was associated originally with meat products. Low-acid canned foods are sensitive products, and great care must be taken to ensure processing procedures resulting in so-called commercial sterility. At present, the most likely source of botulism is from home canned foods that have not been adequately processed and from certain ethnic foods. It is important to note that the perishable canned ham is a candidate for botulism but has a flawless safety record due to the specific protective action of nitrite with the salt (used in curing). These ingredients act against outgrowth of botulinum spores should they be present.

Salmonellosis is an illness that recently has been highly scrutinized by the public. It is probably widespread, and some large outbreaks have been documented. There are numerous serotypes of the *Salmonella* organism, and it occurs commonly in the intestines of domestic animals and birds. This provides a high possibility for contamination of the meat either by contact with intestinal contents during slaughter or by contact with feces clinging to the hide. Estimates of contamination range up to 30% for poultry meat and up to 15% for other meats. The organism is quite heat sensitive so adequate cooking inactivates it. The major problem arises with inadequate heating, or by cross-contamination. For example, by handling raw meat and then transferring the organism to salad contents that are not heated before serving. Salmonellosis usually takes the form of an acute gastroenteritis that may last from one to seven days.

Clostridium perfringens may cause an illness that is often comparatively mild. It usually comes from cooked foods such as meat and gravy that have been cooked and then held warm for an extended period or from reheated leftovers. The cooking heat shocks the spores that then outgrow rapidly during holding. The organism is found in the intestines of animals and poultry. It is not normally a problem in cured meat.

Staphylococcal food poisoning arises from consumption of an enterotoxin produced by *Staphylococcus aureus*. The result is a severe gastroenteritis with an onset usually in 2 to 4 hours, and recovery in 24 to 48 hours. The organism is tolerant to salt and nitrite and therefore a concern in cured meat. It is not a good competitor, but if heating inhibits growth of other microorganisms then it may grow more readily. During growth and production of the heat stable toxin, there is no change in sensory attributes of the food. The organism may be present in animals and is often found in humans in discharge from the nose and throat, and on hair, skin and fingers. Proper hygiene is important, and the organism can be controlled by adequate refrigeration.

Bacillus cereus is usually found in soil, and it is capable of forming spores that may survive cooking. The best control in meat following heating is rapid cooling and proper refrigeration. There are two types of illness, one characterized by diarrhea and the other by vomiting.

Campylobacter jejuni is an important cause of acute bacterial gastroenteritis. It is associated with foods of animal origin, and many domestic animals carry it in their intestines. The organism does not compete well with the other flora commonly present in meat and poultry products. The best control is to prevent cross-contamination and to employ proper cooking procedures.

Enteropathogenic *Escherichia coli* is commonly known as travelers' diarrhea. It is usually contracted from untreated water, and is associated with fecal contamination and poor sanitation. Of recent significance in meat is the illness caused by *E. coli* 0157:H7. One manifestation may be a life-threatening illness characterized by severe abdominal cramps and bloody diarrhea. Animals appear to be a reservoir for the organism and the primary foodstuff implicated is inadequately cooked hamburger.

Listeria monocytogenes has historically been a concern primarily to veterinarians as a cause of abortions and encephalitis in sheep and cattle. In the past few years, however, it has gained notoriety as a food borne pathogen that can occur in meat. The disease it causes in humans may be asymptomatic or may be characterized by septicemia, fever and malaise. In infants, elderly and immunocompromised individuals the mortality rate may reach 30%. The organism is widespread in nature and carried by animals. It can survive long periods under adverse conditions, is salt tolerant and grows at refrigerator temperature. Of greatest concern to the meat industry is cross-contamination to processed products from animals, raw meat, the processing environment or people. Sanitation is the best control.

While the foregoing organisms present the most serious threats regarding microbiological safety of meat, there are three other possibilities to mention. *Shigella* has been of decreasing significance and is associated primarily with seafood. Since transmission from human reservoirs to foods is through improper sanitation, control deals more with food handling than production and processing practices. *Yersinia enterocolitica* may be of a concern since pigs serve as a reservoir, and the organism can grow at refrigerated temperatures. *Vibrio* is responsible not only for cholera, but for gastroenteritis. *Vibrio vulnificus* is a marine bacteria, often associated with oysters and clams, and it is particularly virulent in humans.

VIRUSES

A virus is a piece of nucleic acid surrounded by a protein coat. It is an obligate intracellular parasite, meaning it must be inside a host cell in order to multiply. It is known now that viruses may contaminate foods, usually as a consequence of poor personal hygiene and indirect contamination by food handlers. They cannot multiply in foods and are inactivated by cooking. Therefore, control is made possible by following proper sanitary and heating procedures.

PARASITES

Some parasites may be found in the muscle of domestic meat animal and wild game. Undoubtedly, the most well-known is the encysted larvae of *Trichinella spiralis* that may occur in pork. Ingestion by humans generally results in nausea, vomiting and diarrhea followed by fever and weakness. After the larvae migrate to the muscle and begin encystment, muscular pain and edema may occur. A severe infection may be fatal, but such cases are uncommon. The parasite is inactivated by heating the meat to at least 58.5°C. Certain combinations of frozen temperature and storage time also inactive the larvae. Some countries inspect each individual animal for presence of larvae in the diaphragm muscle.

Cysts of both *Sarcocystis* and *Toxoplasma* may be found in meat and are inactivated by heating to at least 60°C. Tapeworm may occur in muscle of cattle and pigs (*Taenia saginata* and *Taenia solium*, respectively) and can be inactivated by heating to at least 60°C.

CHAPTER 5

PHYSICAL METHODS OF PRESERVATION

Control of temperature (heating or cooling) and regulation of moisture content have been the most widely used means to preserve meat. These have been especially important in the circumstances where chemical preservatives are neither available nor desired. Preservation achieved by control of temperature and moisture is very effective. Cost to control temperature is related to price of energy. The price of energy has fluctuated in recent years, but predictably will continue to increase in the future. While high temperature can be used to kill microorganisms (as in canning), the application of heat or cold and the use of drying techniques are normally employed in an effort only to retard microbial growth. Because of simplicity, these methods (heating meat for example) have been used since earliest times as a means to extend keeping time for foods used in the home. Early advances in technology were made due to needs to provision armies, to distribute food from production areas to more distant population centers and because of a desire to trade food stuffs with other countries. Therein developed the processes of canning and mechanical refrigeration. These obviously remain in very widespread and common use today.

The time-tested methods of heating, cooling and drying have been modified and subjected to subtle variations with the aim of maintaining preservation properties while attempting to improve attractiveness and appeal of the product produced. The result has been development of specific techniques, often in combination with other preservation procedures. Production of a recognized product or branded food has been a sought-after goal.

An example of process modification to produce a new item recognized by many people is the MRE (Meal, Ready to Eat) used to feed troops in the Persian Gulf Conflict. These items basically use the canning or heat sterilization procedure originated many years ago by Appert, except that a flexible and retortable pouch (developed during the 1970's) is used as a container for the entree, and the total meal components, including eating utensils, are packaged uniquely together.

HEATING

Meat is subjected to thermal processing under two broad circumstances—first is production of a pasteurized or commercially sterile meat product and second is cooking before consumption. In the latter case the aim is to improve palatability and appearance, while in the former the additional aspect of stabilizing or preserving the product is also intended. Pasteurized products generally reach an internal temperature of 71°C (160°F), and commercially sterile foods are heated at a high temperature, up to 121°C (250°F), for varying periods of time. To heat a product to commercial sterility in boiling water (100°C) would take a long time, so, as will be described subsequently, the process is conducted at a higher temperature in a pressurized vessel.

Many processed meat products are exposed also to smoke during the thermal processing schedule. Smoke will be discussed separately and as a chemical preservative in Chapter 6.

Heating is a nearly universal event for meat. There are very few instances where fresh meat is not cooked, and processed meats, which have already been heated during manufacture, are very often reheated. Canned meat and meat products represent an important segment of the industry. Approximately 20% of the hog carcass is converted to ham, most of which is made as a semi-perishable canned product. Shelf-stable canned meats include popular items like stew and chili and branded products such as Spam.

Severe heating conditions can result in some destruction of nutrient components, and a decrease in soluble components may result from fluid loss during heating. Overall, however, there is little change in nutrient composition of meat products due to the normally employed heating procedures.

Heat is transferred by conduction, convection and radiation. In the case of conduction, heat moves by direct contact of one particle (of the meat) with another. Convection, however, means a movement or circulation in the substance being heated. Since meat is a more or less solid mass, most of heat transfer occurs by conduction. In a product such as beef stew, conduction is important but convection may also occur.

Heat transfer will occur more rapidly with a greater temperature differential. The composition of the product as well as its geometry are important in determining heat transfer. In solid pieces of meat, the surface will be at a higher temperature for a longer period than the core.

Cooking of meat is done with dry or moist techniques or a combination. The dry procedures employed for home cookery include roasting, broiling, frying and grilling. Moist methods encompass such procedures as stewing, braising, boiling and covered roasting.

Meat products prepared commercially are usually heat-processed either in a smokehouse or waterbath. The smokehouse is really a convection oven designed to deliver certain numbers of air changes per time period, and to control humidity. Provisions are normally present for injection of steam into the smokehouse. The smokehouse system has also been adopted to continuous processing in which product is transported through different zones according to the required conditions of the thermal processing program. Heat transfer to the product is obviously much better in a water bath, and in a retort a much higher temperature can be attained.

Some special situations arise with heating meat. If the meat or meat product is not in a container (can or bottle) or fluid medium then there is evaporative cooling from the surface during heating. As moisture migrates and evaporates from the surface the meat is actually changing in composition. This may affect the transfer of heat, and for example, alters water binding properties that in turn can change the pattern of water migration. Some meat items are now heat processed in flexible bags or pouches. Most sausage products are contained in some type of casing during heat processing. These range from natural casings that are somewhat porous to some synthetics that may be water-impermeable.

The internal or deep temperature continues to rise, when heating of meat is stopped, because the surface layers are at a higher temperature than the core, so equilibration continues. Some meat products are cold showered to "break" the temperature and start it decreasing before the product is moved to the cooler.

A significant item is the time-temperature relation. Both are critical in determining the outcome. The desired goal (of inactivating a certain microorganism) may not be attained, for example, if the temperature is brought up too rapidly or if the endpoint temperature is not held for a sufficiently long period of time. The usefulness of processing records is obvious, as attempting to determine time-temperature conditions after the fact by examining the piece of meat in question is difficult if not impossible.

There are several examples where proper heat application must be verified. Perhaps the best known is assuring that beef, imported from some countries, has been heat-treated sufficiently to inactivate disease-causing viruses such as Hoof and Mouth disease. Another circumstance is assuring that heat treatment has been sufficient to inactivate the parasite *Trichinella spiralis*. Although these are important decisions, the tests available can be considered only qualitative. A simple method used to determine if beef has been heated sufficiently is to squeeze juice from the meat onto a white plate to observe if undenatured red myoglobin is present.

Other tests have been developed to estimate time-temperature conditions to which the meat has been subjected. One is a coagulation test in which the meat is extracted with 0.9% saline and the extract is then heated slowly and observed for protein coagulation. Others are based on determining activity of enzymes such as phosphatase or lactate dehydrogenase. The rationale for these tests is that if the meat has been heated to a sufficiently high temperature for a long enough period of time, then protein denaturation will have occurred and enzyme activity will have decreased accordingly.

Due to the current intense interest in foodborne diseases, there is an even more urgent need for methods to determine time-temperature history of meat products. These will provide assurance in making decisions if an item has been heat processed sufficiently to ensure it is free of various foodborne pathogens.

Time-temperature indicators have been devised to be attached to packages of meat and would indicate if the product has been temperature abused during transit or storage.

Several changes take place simultaneously in meat as it is heated. These are described in the following paragraphs.

Consumers cook meat to improve palatability. As it is heated an aroma is given off and flavors develop. Numerous factors, such as species, age of animal, nutritional background of animal, postmortem aging of the meat and type of cookery, influence aroma and flavor.

Color changes are observed as meat is heated, and these are dependent on the meat being either cured or fresh. The typical pink color of cured meat, due to the pigment nitrosylhemochrome, remains pink as it is heated. In fresh meat, however, the color changes from red to pink to grayish brown. The latter typical cooked meat color is due to denatured globin hemichrome. Browning, primarily of the surface, is another kind of color change and is dependent on method of cookery, the composition of the meat and its previous treatment. While some Maillard reaction occurs (reaction of amino group from protein with aldehyde or ketone of reducing sugars to produce a brown color), other reactions such as charring also take place.

Surface browning, from a culinary viewpoint, is desirable, but there is, on the other hand, some concern about formation of mutagens being associated with the reaction products. However, another proposition is that meat contains naturally occurring anti-mutagenic compounds.

As meat is heated, it loses weight and the composition is altered (unless it is in a sealed container). The weight loss is primarily due to evaporative moisture loss and is usually greater at a higher cooking temperature. At some point a dehydration or hardening effect may also occur at the surface, which will retard further evaporative loss. When moisture content decreases the concentration of other components increases, unless they are soluble and have been carried away to the surface. Lipid is also involved because as the temperature rises the lipid becomes more liquid or fluid, and if fat cell membranes rupture the lipid released may migrate within the structure of the muscle or to the surface and be lost as drip. The liquified lipid distributed throughout the muscle, undoubtedly gives rise to desirable palatability properties.

As temperature rises, denaturation of the proteins takes place. Some of the muscle proteins begin to denature as the temperature reaches 50°C, and most of the denaturation occurs in the temperature range 60 to 70°C. The proteins become less soluble as denaturation proceeds, and this is important, for example, in stabilization of processed meat emulsions. Water binding ability of the proteins is decreased, so water may be more easily lost.

As a result of denaturation of protein, the texture becomes firmer. Textural changes are related to loss of water and what is perceived as a drying of the meat.

A protein of considerable importance in muscle is collagen. It is quite pervasive surrounding each individual fiber and occurs also in thicker layers around the fascicular structure. It is sensitive to temperature and undergoes a shortening or shrinkage in the temperature range 60 to 70°C, being subsequently converted to gelatin. Thus, heating alters collagen and improves tenderness.

CANNING

The canning industry is based on the amazing and original efforts of the French worker Appert that appeared in print in 1810. Long before an understanding of the bacteriology involved, he worked out a method to preserve food by sterilizing it with heat in a hermetically sealed container. Before his discovery, the only methods to preserve food for extended periods were by drying or adding a substance such as salt.

The procedures he stated are quite simple and consisted of: (1) enclose the food to be preserved in a bottle or jar, (2) cork with great care—success depends on this closing, and (3) submit to boiling water.

The basics of the original method are still used extensively in the meat industry today. The aim is to destroy spoilage and toxigenic microorganisms while simultaneously maintaining acceptable quality and textural properties in the meat. This is best accomplished by exposing the meat to a high temperature for a short time.

Spores are generally quite heat resistant and represent the major problem to solve in preparing canned meat. The commercially sterile products should be stable at ambient temperatures for several years.

One type of canned meat item requires special mention. It is the canned ham, and it is not stable at ambient temperature as are the usual "canned" items. These pasteurized products usually reach a temperature of at least 65.5°C, and depend on salt and nitrite to prevent outgrowth of spores (not destroyed by the low heat processing) should they be present and should the product be temperature abused. The typical product made in this manner must be labeled with the statement: "Perishable—Keep Under Refrigeration." Under proper refrigeration they should maintain high quality for two years.

Even in a boiling water bath, as used originally by Appert, at 100°C it would take extended time periods for a large piece of meat to reach that temperature, and obviously, it would be impossible to reach higher temperatures. Attempts were made to increase the boiling point of the water by adding salts. But, the real solution came with development of pressurized vessels (called autoclaves or retorts) in which steam could be injected under pressure to attain temperatures much higher than the boiling point of water. The retort may be non-agitating, or in another type the cans are moved so that convective heating occurs. In the so-called hydrostatic retort, cans are moved through a combination of water and steam chambers.

Meats are usually heat processed in tin, glass, aluminum or plastic containers of varying shapes including round, pear-shaped and square. In recent years much more interest has been shown in flexible types of pouches. Many hams or ham products are now manufactured by a so-called "cook in bag" process.

The ordinary tin can is actually steel with a thin coating of tin. In addition, there may be various coatings on the interior depending on the food application intended. In the case of meat it is usual to coat the lining with a sulfur-resistant material. Without this coating sulfur compounds released from the meat by heating would stain the tin a black color. Easy release or non-sticking compounds may also be used.

Determining a heat processing schedule for a given meat product is an essential, critical and complicated process, and standard texts should be consulted for details. The plan is to inactivate the most resistant spores that could possibly be present. For the specific product and container in question, heat penetration data, using the equipment to be employed, must be obtained. This usually means doing temperature profiles in the geometric center of the contained product using the projected thermal processing cycle. Information must be known about the thermal death time of the most resistant organism projected to be present and, finally, appropriate calculations must be made.

The possibility also exists for using microwave technology in a heat sterilization procedure. The advantage is to produce a high quality food due to the rapid rise in temperature over a shorter time course.

An alternative to the typical canning process is aseptic processing. In this process the product is heated, allowed to cool and then transferred to a sterilized container and sealed—this all being done aseptically and without contamination. The major application is with fluid foods, which can be heated at a very high temperature and short time while exposed to a heat exchanger. Due to the short time-temperature exposure, the result should be an extremely high quality product. Obviously, this technique would not work with solid pieces of meat, or with fluid products containing pieces or chunks.

Another recent innovation is called the "sous vide" technology. Food pieces of similar size and thickness are arranged and sealed under vacuum in impermeable plastic pouches. The pouches are cooked at relatively low temperature in a circulating water bath for up to four hours, and then chilled and held refrigerated for up to three weeks. The preparation is reheated in the same package before serving. The slow cooking under vacuum is claimed to tenderize the meat and retain flavor and nutrients; the result is attractive, palatable and of very high quality. The package (vacuum) is pasteurized but not sterilized. Therein lies the potential problem with the possible growth of pathogenic, anaerobic organisms should the package be temperature abused. The safeguards include use of only high quality raw products, strict hygiene, tight control of process and attempted assurance that the product will be handled properly until prepared and consumed.

Extrusion of a food produces an increase of temperature in the product. It is mentioned as a potential, specialized process that does subject the food to increased temperature.

COOLING

In contrast to heating, which is used to inactivate or at least severely impair functioning microorganisms, cooling techniques are employed mainly to slow or limit their growth. Two levels of cooling are used—refrigeration and freezing. Refrigeration is the most widely used and effective means of preserving meat. It is employed at the slaughter plant immediately after slaughter, during transport and storage, and at retail display.

Freezing is a more effective means to preserve meat over an extended time period. Although freezing is so effective and is used extensively for storing meat, it has not been adopted for retail use with meat as it has with other food items such as vegetables.

The several attempts to market frozen meat have been unsuccessful due to consumer resistance. These same consumers will purchase fresh meat and then freeze it at home under very much less than ideal conditions. Apparently, the consumer has some aversion to purchase of frozen meat probably thinking it has been frozen to hide a defect. Appearance has also been a factor. The consumer wants to see the meat surface, especially the color. Therefore, frozen meat cannot be sold in opaque packages. Early attempts to make retail frozen packages in transparent film had a problem with the meat color being dark. In some instances frost pockets also became visible.

The lower the temperature of refrigerated storage the slower is microbial growth (as well as enzymatic activity). Therefore, efforts are made to keep meat at as low a temperature as possible within the constraints of equipment performance and economics of operation.

Meat freezes at about −2°C (28°F). Another possibility is to hold meat at a "just frozen" temperature of −5°C to obtain good preservation yet minimize energy cost.

Both refrigeration and freezing are excellent means of preserving the freshlike appearance of meat. It must be remembered, however, that neither can be used to improve quality of meat.

For refrigerated and properly packaged retail meat, the shelf-life is probably 72 hours, after which some discoloration can be expected to appear. Ground meat may display acceptable condition for only one day. For vacuum packaged and refrigerated meat the quality is expected to remain stable for at least three weeks. In the case of cured meat, that is vacuum packaged and refrigerated, at least two months is expected before obvious changes become visible. Frozen meat keeps well for up to a year. Understandably, the above times are greatly influenced by species, level of hygiene practiced and performance of equipment used.

The initial refrigeration of meat occurs as carcasses are moved into the chilling rooms, and the desired results are achieved by adjusting and balancing several factors. Temperature, air flow and humidity in the chilling rooms are important factors, as are size, shape, composition and uniformity of the carcasses moved into them. Because of the large size and irregular shape, lowering the deep temperature of a beef carcass fairly rapidly is an interesting engineering problem. A typical cooling curve is shown in Fig. 5.1, and illustrates the several hours required to chill a carcass. The difference between beef and pork is given.

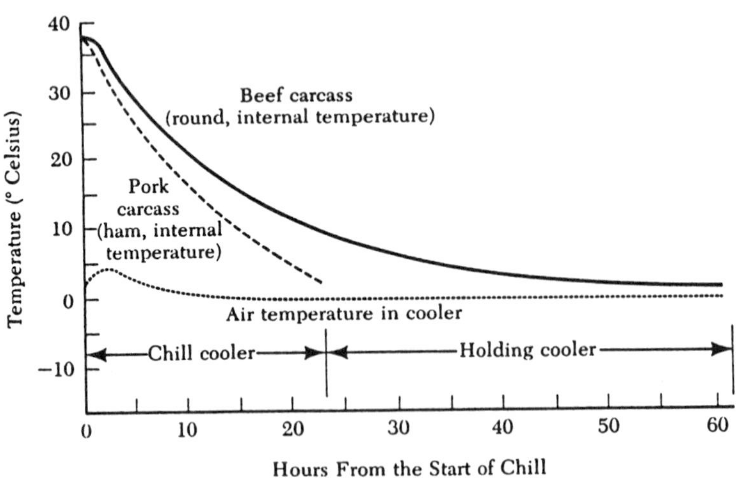

FIG. 5.1. COOLING CURVES FOR BEEF AND PORK CARCASSES
From R. Retrum. 1957. Adapted from Fig. 10.1 and 10.2 *In*: The Science of Meat and Meat Products, 3rd Ed., (J.F. Price and B.S. Schweigert, eds.) Food & Nutrition Press, Inc., Trumbull, CT.

Carcass chilling rooms are normally operated in the temperature range of −2°C to −4°C (28° to 25°F) with a relative humidity of 88 to 92%. Shrinkage loss from the carcass is in the range of 1 to 2%. A lower temperature differential gives more rapid cooling. The ability of the equipment to maintain a constant low temperature as warm carcasses are moved in is important. Air movement promotes more rapid chilling—as well as greater dehydration at the surface. In general, adjusting conditions for more rapid cooling results in greater weight loss of the carcass. If cooling does not occur rapidly then the potential exists for greater microbial growth and subsequent shelf-life may be shortened.

Standards in the European Community call for the deep temperature of beef to reach 7°C before cutting or shipping (in reality, in about 18 hours).

In the case of poultry, chilling is accomplished rapidly by immersing the carcasses in an ice-water bath. Besides rapid chilling due to the direct contact with the chilling medium, other considerations are weight gain from water binding, some lightening of color due to surface extraction of pigment by the water and the need to control microbiology in the bath.

The export of meat from Australia to Great Britain from about 1900 onward is a historical chapter that provides examples of how special problems in preservation were solved. Transit by sea was about 55 days, so microbial spoilage was a real threat. Shipment of frozen beef was discriminated against because of the severe drip problem encountered when it was thawed. Therefore, a great deal of scientific research was conducted, which greatly benefitted the industry and resulted in development of some useful and scientifically grounded procedures.

Attempts were made to extend refrigerated storage time by exposure of the carcasses to formaldehyde—this was deemed unsuitable because of possible negative effects on human health. The so-called "Bullot Process" used the combustion products from sulphur, potassium nitrate, wattle bark and charcoal, but it also was unsuccessful. Various combinations of air circulation and humidity control in the holds of the ships were also attempted.

The Australians did make successful sea shipments of chilled beef carcasses during the early 1930's by adding a 10% CO_2 atmosphere to the holding rooms. Carbon dioxide was known to be an effective microbial inhibitor. The process was successful because it also incorporated strict hygienic standards at all stages, such as animal holding, slaughter, chilling and transportation. Their continuing success today in terms of excellent storage time for chilled and frozen meat is also a reflection of strict hygienic standards.

In recent years, there has been increased interest in the techniques of accelerated cooling. In most instances this is accomplished by using equipment capable of generating and maintaining extremely low temperatures such as in the range of −15 to −35°C or by spraying with or immersion in cryogenic liquids.

Liquid nitrogen is the cryogenic substance of choice and in some cases solid CO_2 or dry ice may be utilized.

In such cryogenic procedures, the surface layer of the carcass or meat is usually "crust frozen", and equilibration is then allowed to proceed under normal refrigeration. Obviously, precautions must be taken so as not to induce cold shortening and associated toughening, or pre-rigor freezing and the associated drip problem. Accelerated chilling is expensive and so therefore the procedure must produce an advantage by giving higher quality meat or by increasing yield. Since the surface is actually frozen there is a barrier to evaporation of water and the yield is improved.

FREEZING

Freezing provides an excellent means for storing meat for long periods of time while maintaining it at a high level of quality. Meat is a food substance that freezes easily and well. As is always the case, quality cannot be improved once the process is initiated. It must be recognized that freezing does not normally kill microorganisms. Similarly, viruses are not deactivated. However, frozen storage at certain time-temperature combinations is effective in destroying parasites such as trichinae.

Although frozen storage of meat is extremely important, there has been and continues to be strong consumer resistance to retail purchase of frozen meat. On the other hand, individual meat portions for use in the institutional trade are often distributed in the frozen state. Much of the meat for military use is distributed and stored frozen, and the meat industry itself makes extensive use of frozen meat for inventory control.

Although cured meats can be frozen, the freezing process is not as often used as it is with fresh meat. Cured meat can be stored successfully and longer under refrigeration than can fresh meat. However, cured meat is more susceptible to oxidative and textural changes when held frozen. Cooked meats may be frozen, but development of off-flavors may be a problem.

The recommended storage temperature for frozen meat is −18°C (0°F). A lower temperature may be more desirable in maintaining quality but, of course, the advantage gained must be balanced against additional cost.

There are four areas or phases of importance in a consideration of frozen meat; these are pre-freezing, freezing, frozen storage and thawing.

Pre-freezing relates to the quality of the meat and to preparing it for freezing. As mentioned already, the quality of the frozen product will be no better than the starting quality of the raw meat—this includes both the intrinsic quality of the meat (factors such as tenderness and flavor) and the microbiological situation. The time postmortem that freezing is undertaken plays a role. If temperature is decreased pre-rigor with sufficient rapidity, then cold shortening and associated toughening may result. The rate and extent of postmortem change and the time postmortem can be quite important since pH and microstructural condition are related, for example, to drip.

Other important pre-freezing considerations are size and shape of an item to be frozen and how it is packaged. Under given conditions, slower freezing will occur with a larger size or a cube shape compared to a unit that has one thin dimension. Most effort is now directed at boneless meat, since irregular shapes and sharp edges of bones cause special problems. Impermeable packaging is important to protect the surface of the meat from exposure to oxygen and from water loss during freezing and storage. Packaging should be applied in a manner to minimize dead space within the package that would also allow migration of water from the product.

The actual freezing is accomplished by exposing the meat to a temperature differential that can be created by numerous combinations of factors depending on the desired outcome and the cost of doing it. Still air can be used to freeze meat, but the process is greatly speeded by increasing air flow as in a blast freezer. Actual contact of the meat with a plate type device is faster still and most rapid is use of a cryogenic agent. The process can be batch type or continuous.

Typical freezing curves are shown in Fig. 5.2 and 5.3. In the classical circumstance illustrated in Fig. 5.2, there is an initial decline in temperature with a supercooling shown in which the temperature actually falls below the freezing point of the meat. When ice crystals are formed, the heat of crystalliza tion produces a temperature rise up to the freezing point. The temperature then declines. This is slower at first, as there is more water available for crystallization during which process heat is generated. The percent water frozen at a given temperature is shown during the declining temperature phase.

Figure 5.3 illustrates the circumstance where warm beef carcasses are moved directly to a blast freezer. The effect of a more rapid air movement is clearly indicated, especially in terms of the time required to reach the final temperature. The comparison of total carcass with a small, thin piece of meat is also shown.

The general consensus is that rapid freezing produces smaller ice crystals than slow freezing. If the meat is post-rigor and the freezing is relatively slow, the ice crystals probably form in the extracellular spaces. If the meat has not yet gone through rigor or if the freezing rate is quite fast, then the crystals probably form within the fibers. In the latter case the structure of the fibers is such that the growth of the ice crystals is limited by the cell walls and results in longitudinal spears of ice.

Rapid freezing that causes the formation of small crystals on and near the surface of the meat provides a physical situation that results in a reflection of light that gives the impression of a lighter colored surface.

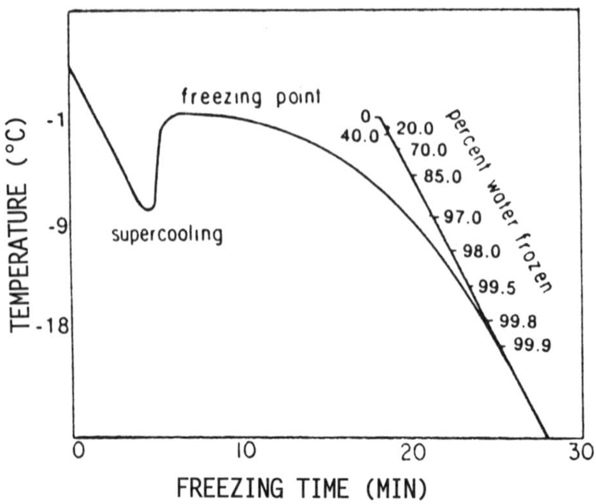

FIG. 5.2. FREEZING CURVE FOR THIN SECTIONS OF BEEF
From N.W. Desrosier and J.N. Desrosier. 1977. The Technology of Food Preservation, 4th Ed., Van Nostrand Reinhold/AVI, New York. Reproduced by permission of Van Nostrand Reinhold/AVI, Inc.

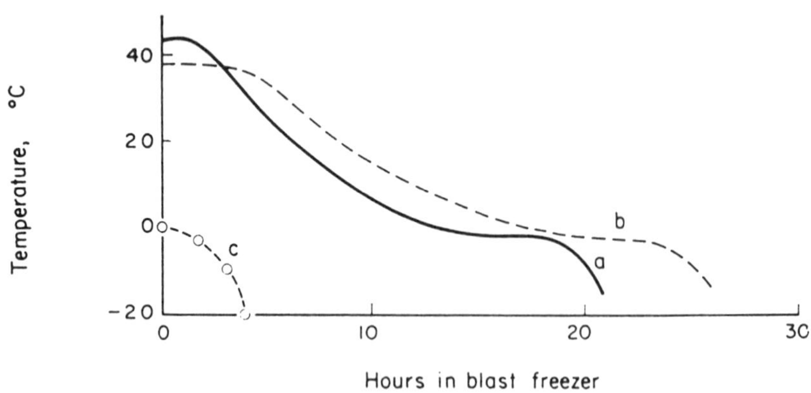

FIG. 5.3. TIME-TEMPERATURE CURVES FOR WARM BEEF PLACED INTO A BLAST FREEZER
(a) Temperature in deepest portion of side: blast at 1,000 ft/min, −40°C. (b) Temperature in deepest portion of side: blast at 250 ft/min, −30°C. (c) Temperature at center of thick section of sirloin. From R.A. Lawrie. 1966. Meat Science. Pergamon Press, New York. Reproduced by permission of Pergamon Press.

The phase that is potentially the most damaging to the quality of the meat is that of frozen storage—especially if there is fluctuation of temperature. This is the time when growth of ice crystals can occur and result in structural damage. There may be recrystallization and movement of water.

Freezer burn occurs when moisture evaporates from the surface of the meat. This normally takes place as a result of poor packaging so there are air pockets in the package. The outcome is a blanched color and a more or less dehydrated layer on the surface of the meat.

Even though low temperature slows enzymatic and chemical reactions they may still proceed in frozen meat. For example, oxidation of lipids occurs during frozen storage of meat. There is also a zone during the freezing process where reactions may proceed rapidly. As freezing progresses and water is being crystallized, the concentration of solutes actually increases, which may speed up reactions. This could be especially important if freezing rate is slow and high solute concentration is present for longer time periods.

The final phase to consider is thawing. Thawing is essentially the reverse of freezing, but as shown in Fig. 5.4 there is a major difference and that is that the block of meat is near the freezing point for an extended period of time. This happens because normally it is not possible to use the same degree of temperature differential for thawing as is used for freezing. The situation is intensified in larger blocks of meat. The problem is that the surface of the meat may be at ambient temperature for a very long time period while the center of the meat remains frozen. Since most of the bacterial contamination is on the surface of the meat, the situation provides an excellent condition for growth of microorganisms.

Thawing may be done in cold or warm air; a flow or movement of air speeds the process. Immersion in water gives an improved thawing rate due to the direct contact, but the meat must be in an impermeable bag or there will be surface extraction of meat constituents such as protein, and the problem of microbial cross-contamination also exists. Use of microwave equipment, so that heating occurs from the inside also, appears to be an applicable way to thaw meat. Problems relate, however, to uneven heating and the equipment requirements for the relatively large blocks of meat routinely used in the industry. Tempering, or bringing the temperature up some from that of frozen storage, is possible with microwave equipment.

Under industrial conditions the problem of thawing is mainly one of logistics. That means not only having relatively large amounts of space to thaw meat but also being able to predict how much should be thawed.

Flakers may be used to shave pieces off frozen blocks. Normally, this requires some prior tempering or bringing the temperature of the frozen meat up some from the usual frozen storage temperature.

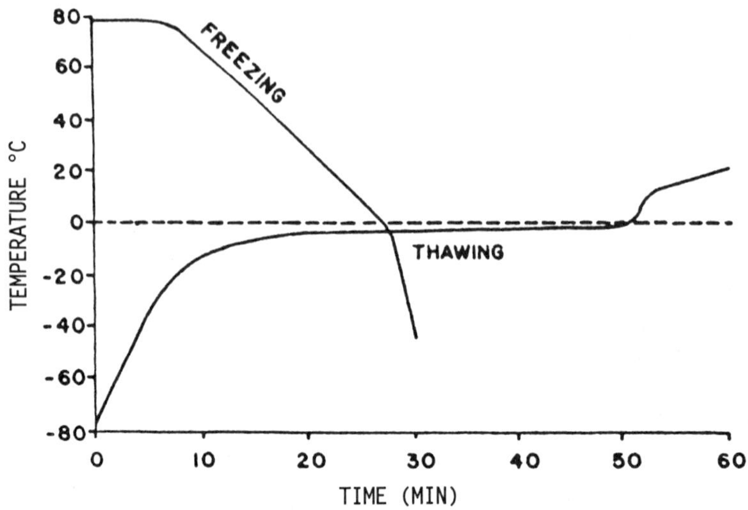

FIG. 5.4. COMPARATIVE FREEZING AND THAWING CURVES FOR THE GEOMETRIC CENTER OF A CYLINDRICAL SPECIMEN

From O. Fennema and W.D. Powrie. 1964. Fundamentals of Low-Temperature Food Preservation. Adv. Food Res. *13*, 219. Reproduced by permission of Academic Press, Inc.

The phenomenon of drip may be experienced. Drip is a watery, red exudate that is not only unsightly but also contains nutrients such as proteins, vitamins and minerals. Obviously, drip results in a shrinkage or weight loss. Drip was probably more of a problem when whole carcasses or quarters were frozen for shipment, as explained earlier in this chapter. It is most severe if the pH of the meat is low, giving poor water holding properties. An amount of exposed, cut surface of the muscle is necessary for the drip to be expressed.

CONTROL OF MOISTURE

Drying as a means of preservation is a very old technique, and basically is accomplished by removing water. If water is removed from meat a point is reached where the available water is not sufficient to support microbial growth. As the amount of water is decreased other substances are actually increased in concentration in the smaller amount of remaining water. Certain foods, such as grains and nuts, become adequately dry on the plant so that they can be stored for years.

Some early societies dried meat by exposing strips of it to sunlight. In some cases heat was applied and in others, such as with fish, light salting was used in addition to the drying. With simple air drying, the control of proper hygiene may be a problem because the meat is exposed to ambient conditions such as dust, insects and animals.

One percent or less of the total meat production today is converted to true dried products. The biggest segment is manufacture of dehydrated flakes and pieces of chicken and beef that are used, for example, in dry soup mixes.

Extremely lean meat is the most desired for drying. Fat is subject to oxidation and the production of off-flavors, especially if heat is used in the drying procedure. The use of an antioxidant may be nearly prerequisite to production of a good product. Denaturation of protein and some loss of nutritional value may occur, and depends largely on the heating process. This can markedly affect rehydration. Dried products often have a wrinkled surface and may have voids within the internal structure, which result from the evaporation of water.

In contrast to simply drying in the sun, most preparation of dried foods is more correctly referred to as dehydration in which there is a least some degree of control over air temperature and movement, and control of humidity. In fact, some of the dehydration systems are quite sophisticated. Driers range from simple boxes to continuous systems. Other systems, such as spray drying, have special applications for certain food forms.

Dehydration of meat often involves a cooking step before the drying. In this case the beginning moisture content is about 50% and is lowered during the process to about 4%. This is contrasted to a moisture content of 25 to 50% in typical dry and semi-dry meat products (see Chapter 7 for a discussion of these products).

Production techniques for dry sausages illustrate the care that must be taken regarding control of temperature, humidity and air flow. From the interior of the product, water moves by diffusion to the surface where it is then removed by evaporation. The conditions must be properly balanced. If moisture is allowed to evaporate too rapidly from the surface, a hardened layer (called "case hardening") will develop and make it difficult for water remaining in the interior to migrate through to the surface. If surface water is not evaporated rapidly enough it will collect and allow the possibility for growth of undesirable microorganisms.

Pemmican is a unique type of dried, meat-containing food. It was made by early Native Americans and used as an emergency or travel ration. Sun-dried or heat-dried lean buffalo or venison was pounded to shreds or powder, mixed with hot melted fat and sealed in hide containers. Subsequent variations were made by adding berries, fruits, nuts or other food substances. The Armour Co.

manufactured several types of pemmican that were used, for example, in the polar explorations.

Some typical foods such as dates, jams and country ham are intermediate moisture, meaning they contain 10 to 40% water and have an a_w of from 0.65 to 0.9. Interest has been evident in manufacturing a special class of intermediate moisture foods by adding a humectant to lower a_w. An antimicrobial, and possibly other chemicals, are added to maintain stability and palatability. Intermediate moisture foods have been used successfully in the pet food industry, but hardly used at all for human food, except for some space flights and specific military application.

Humectants typically used include sugar, salt and glycerol. These can be incorporated by soaking a solid piece of food until sufficient exchange has taken place to lower the water activity to the desired point. Alternatively, the soaking can be done to a piece that has been dehydrated first so as to make the exchange easier. Another possibility is to blend, cook and extrude the ingredients. A problem to overcome is taste imparted by the humectant—for example, the sweet flavor may be overpowering to some and is difficult to mask.

Freeze drying is another method of preservation by moisture control. It was instituted first in the early 1900's as a means for high quality preservation of biologicals such as human serum. It is a means to remove water by sublimation. In other words, water evaporates from the ice without the ice melting. The process is accomplished by freezing the food substance and then subjecting it to properly controlled vacuum and temperature conditions.

The advantages are that there is little or no shrinkage or distortion, and the structure of the meat remains in excellent condition. The results, compared to dehydrated food, are greatly improved rehydration properties, as well as excellent nutrient and flavor retention. Whole steaks and chops can be freeze dried.

A typical process might take 12 hours or more. The frozen product is subjected, ideally, to a pressure of 1 mm Hg and is placed on a tray or plate that is in fact a heating unit. The key to temperature control is to heat it enough to drive sublimation along without melting the ice. A typical process may be at 43°C, for example. In meat, the water content is approximately 6% by the time the ice has sublimed. The goal is to get the water content finally down to 1%, and this may take as much time as did the initial removal of water. Since the ice is removed at this point there is no chance of melting it and so the temperature can be increased to speed the final removal of water. The limiting factor is to not go so high as to scorch or otherwise damage the food structure or its nutrient content. Microwave heating can be used to greatly speed the process.

The result of a properly done process is a porous, sponge-like structure. This provides for a rapid and good rehydration but it also allows easy penetration by atmospheric moisture and oxygen. Therefore, the finished freeze dried product must be adequately protected by proper packaging. With adequate packaging, the product stores extremely well.

The cost of freeze drying is generally two to five times higher than if an equal amount of water had been removed by conventional means. Freeze dried meat has application for military use, for campers and explorers and is also used in some dried food mixes such as soups.

MICROWAVE APPLICATION

Microwave technology has possible application in numerous processing procedures, as well as in home preparation. It is used industrially, often in combination with other conventional heating procedures, in such operations as cooking, tempering, dehydrating, pasteurizing, sterilizing and freeze drying. The microwave component provides a rapid rate of internal heating while the conventional heating is used for surface effect or to assist in temperature equilibration.

The very early use of microwaves was as a heat therapy that by the 1920's was known as diathermy. Application in food processing was under consideration in the early 1940's, but engineering efforts were directed, obviously, at war efforts such as development and refinement of radar technology. Papers on food applications began to appear in the late 1940's as did commercially available microwave ovens.

Microwaves fall between radio waves and infrared radiations being from 0.025 to 0.75 micrometers in length. These correspond, respectively, to frequencies of about 20,000 and about 915 megahertz. The most commonly used frequencies in food applications are 2450 and 915 megahertz, which have penetration depths in foods of about 30 cm and 10 cm, respectively.

Microwaves travel in straight lines, are reflected by metals and for the most part pass through such substances as glass, paper and plastic. They are absorbed by food constituents including water. When microwaves are reflected, the reflecting surface is not heated. When they are absorbed, however, they cause the absorbing substance to become heated. In thus heating a substance they lose energy (called "loss") within the material being heated.

When microwaves enter into a food the polar molecules of the food tend to align, but the microwaves are reversing so many times per second that the molecules oscillate. The oscillation is so rapid that it more or less generates an intermolecular friction causing the food to heat. Different components do not necessarily heat uniformly, but as the process continues conduction evens out the potential differences.

Conventional heating is a surface inward effect, meaning that the surface is exposed to a higher heat for a longer time. In the case of microwave heating, penetration to the interior of the food occurs and heat is generated quickly and uniformly throughout.

Microwave technology has many potential industrial applications—some of which are now in use. It seems, however, that overall the procedures have been somewhat slow in being adapted and utilized in industrial processing of foods. While it is difficult to provide an explanation for this conclusion it may be explained in part by a slowness of engineering design for food application.

Microwave heating is 10 to 20 times faster than that of usual thermal transfer. With microwave techniques, cooking time is linearly related to quantity of food.

Heating rate depends on the shape, size and structure of the substance being heated. Low moisture products have greater penetration depths and can therefore be thicker. Heating of a given product may be uneven; therefore, use is made of wave stirrers and rotation of the product in the oven. Some methods incorporate other conventional methods of heating as part of the total procedure.

Two other considerations to be dealt with are the fact that surface browning does not occur or is minimal, and the possibility of low microbial inactivation because heat is not concentrated at the surface where the greatest concentration of microorganisms occurs. Microbial inactivation is not due to the microwaves themselves, but instead to the thermal change caused by them that in turn denatures proteins and nucleic acids.

The benefits from using microwave heating in industrial application are to accelerate the process time, to improve quality, to reduce costs and to increase yield. Conveyorized or continuous systems have improved industrial microwave use compared to batch type ovens. At present, the technique is being used for pre-cooking bacon, chicken parts and meat patties. Procedures are available to use microwave techniques to set frankfurter emulsions—a process that would eliminate need to use and remove casings.

The technique is also used to thaw, or more correctly, to temper frozen product. For whole fish and meat blocks, tempering raises the temperature but keeps it below the freezing point so that the product is still frozen but not solidly so.

Home use of microwave ovens has increased dramatically with 70% of homes in the United States now having one. Engineering efforts and changing demands by the consumer have resulted in a change from the early home ovens, which were large and expensive, to today's models that are compact and relatively inexpensive.

The interest by consumers in the convenience and speed associated with microwave cookery has stimulated the industry to make major advances in the types and availability of suitable products. For example, there are now available shelf-stable, prepared entrees and refrigerated, pre-cooked entrees. Microwave ovens generally consume less energy for preparation of a single meal item than does the same preparation in a conventional electric oven.

Uneven heating problems may be experienced in meat that contains bone. It was also discovered that trichinae could sometimes survive microwave cooking procedures that brought the meat to a correct end point temperature. This was undoubtedly due to the rapid heating rate that meant exposure to a given temperature for a shorter period of time. This can be corrected easily by holding the product for a time when it comes out of the oven so that the heat will be distributed uniformly.

IRRADIATION

Irradiation, as a process to pasteurize or sterilize meat, has enormous potential. It has been an available procedure for a third of a century, yet has not been well or widely used. This has been due almost entirely to fear that the consumer will reject irradiated foods on the basis that they may be unsafe.

Private development work began in the 1940's, and in the 1950's the work on sources, equipment and wholesomeness of the product produced was of major magnitude and was sponsored primarily by the government—not only in the USA but also in other countries. This was due to the fact that human use of irradiated foods would have to be approved by government agencies.

When ionizing radiation is absorbed by a biological substance it leads to breakage of chemical bonds within molecules and the formation of new substances. It produces ions and other activated molecules, which react further. In other words it causes a chemical change. If the energy level used is sufficiently high the process may induce radioactivity in the target substance. Obviously, this is unacceptable in food. To avoid this the energy level of gamma rays or X-rays should not exceed 5 MeV and that of an electron source should not exceed 10 MeV.

The sought-after application with food is the inactivation of bacteria, yeasts, molds and parasites. Mode of action is by altering the DNA that damages biological function resulting finally in death of the organism.

The primary consideration in irradiation of foods is the dose or amount delivered. Higher dose levels effectively kill organisms resulting in a product, which is indefinitely stable even without refrigeration. This sterilization process is termed radappertization. A lower dose can be used merely to reduce the number of microorganisms present so that the product then lasts longer under refrigerated conditions. This is termed radurization.

The dose required for sterilization of a given product is determined in a manner similar to thermal processing schedules. Irradiation does not, in general and compared to thermal processing, decrease nutrient content. Also, water-binding is not affected and the texture of the meat remains well preserved.

Irradiation is also known as cold sterilization. It gives the same preservation properties as a thermal process but does not raise the temperature of the product.

There are other factors to consider. High doses may produce an undesirable, scorched flavor. Early on in development the off odor or flavor was described as "wet dog." The flavor problem can be minimized by conducting the irradiation process at lowered temperature. In fact, it is often conducted on frozen meat. The usual procedure for radappertization of meat is to heat process first followed by irradiation, since enzymes are not inactivated by the irradiation.

In the case of cured meat, radappertization can be used to control the potential for outgrowth of *Clostridium botulinum* spores should they be present. This permits a much lower level of nitrite use—in fact, at a level only for development of typical color and flavor. The end result is less potential for formation of nitrosamines in the cured meat. Irradiated ham has been prepared for and used successfully in some of the space programs.

Radurization has potential as a means to extend shelf-life of the perishable fresh retail meats. The typical retail meat package is prepared on site and usually does not maintain suitable appearance for more than three days. Radurization could be used to lower the level of microorganisms and obtain longer shelf-life, or perhaps move toward a centralized prepacking system. It must be remembered, however, that changes such as color alteration, lipid oxidation and drip can proceed without the presence of microorganisms.

Radurization probably will not find application with processed meats. The combination of salt, nitrite and vacuum packaging provides extended shelf-life.

Other potential applications would be to control pathogens such as salmonellae in poultry and parasites such as *Trichinella spiralis* in pork.

The factors of proving wholesomeness, labeling, and consumer fear have greatly slowed commercial use of irradiation in the food processing industry.

Since 1963 the United States has allowed some limited use of irradiation, such as controlling insects in grain and preventing sprouting of potatoes. In 1981 United Nations organizations announced that any food irradiated up to an average dose of 1 Mrad or less is wholesome for human consumption and therefore should be approved. Since 1984 other dose-specified uses have been allowed, such as for disinfecting spices. In 1984 use to reduce risk of trichinosis in pork was permitted, with proper labeling of the resulting product.

In January 1992, a plant in Florida was ready to begin shipping irradiated strawberries amid much opposition from consumer advocates. In September 1992, the irradiation of poultry products was approved. The regulation allows the treatment of fresh or frozen, uncooked whole poultry carcasses or parts. It remains to be seen how the use of irradiation in foods is accepted by consumers.

ULTRASOUND, HIGH PRESSURE AND HIGH-VOLTAGE PULSES

Ultrasound (ultra short waves of high frequency) can be used to destroy microorganisms in a liquid medium. The technique has not found application with meat, probably because of the possibility of inflicting damage to the food item itself.

Similarly and under certain conditions, high pressure and high-voltage pulses can inactivate microorganisms. Some use may be found for these procedures in the process of food preservation in the future.

PACKAGING

Packaging is discussed at this point because it is critical to maintain proper stability after the preservation procedures have been applied to the meat. Packaging is an integral part of the total preservation scheme. Very major advances have been made and continue to be made in packaging science. The work has depended on a knowledge and understanding of the biology of meat, expertise in polymer science as related to continuing developments in flexible packaging and engineering aspects of equipment design.

Canned meat is thermally processed within the package (can). This idea has been extended to "cook-in" products, such as hams, which are thermally processed within a flexible film container. Developments are appearing rapidly wherein meat items are made available at retail in consumer-sized containers, which are heated directly for preparation.

Besides being a container in which processing can be done, packaging is most often something used to enclose the meat in order to protect the desirable properties inherent in or imparted to it. This means, for example, that the aspect of color should be clearly displayed to the consumer in the desirable and expected state, while simultaneously protecting the product from moisture loss and contamination by microorganisms. The goal is to stabilize what has been achieved in preparation or processing, and because meat continues to be an active biological system, even after packaging, many challenges and possibilities are provided.

The package is often used as a container for distribution of the product. Packaging also offers a place for color, graphics, form or shape and other features that are used for product identification. The recognition factor is extremely important in promotional and marketing schemes. A normal part of packaging is a label that conforms to government approval regarding such factors as ingredients and weight. The package also provides space for supplemental ideas such as nutritional information. Of increasing importance is the concept that handling precautions and preparation guidelines may be transmitted to consumers via the package label.

The idea of vacuum packaging is a milestone of advancement in the meat industry. Also, the development of variety and specific function in flexible films has been enormously important. One need only contrast the situation today with that of little more than a half-century ago when meat was distributed in cans, in carcass form or in butcher-wrap paper. Today, meat packaging films and containers are essentially custom made for individual application. Many are multilayer and made for certain clarity, barrier, strength, machinability and sealing properties.

There are several methods of applying the packaging material. With fresh meat it is simply an overwrap, meaning the film is wrapped over the meat piece and then heat sealed. Preformed bags or pouches may be used. High speed sophisticated machines use roll stock or webs of film whereby the film is formed around the product or a die forms a container for it. The possibility of utilizing vacuum, gas back-filling or heat shrinking are also available.

In recent years a somewhat negative aspect has arisen regarding the advanced technology of packaging, and that is a charge of overpackaging. There are environmental concerns, that an enormous amount of nonbiodegradable material is being produced. This charge is often leveled at the food industry; in fact, the food industry produces only a fraction of the total in comparison to the packaging used on all consumer goods. There is also concern about migration of packaging components (chemicals) to the food they enclose, especially when the package is used as a container for heating.

For fresh meat the packaging is designed to allow oxygen transmission, so that the bright red color of oxymyglobin is produced and maintained, and to retard moisture loss so the product does not dehydrate and lose weight. Fresh retail cuts are typically prepared at the store, overwrapped and heat-sealed with a film such as polyvinyl chloride. Shelf-life is normally three to five days. Proper refrigeration and hygienic measures are important. Centralized prepackaging of fresh meat has been slow to develop.

One way to capture much greater product stability in fresh meat is to vacuum package it—this provides at least three weeks time under adequate refrigeration after which the meat can be exposed to air (oxygen) and it will bloom to a bright red color. However, while the meat is in the vacuum package myoglobin remains in the reduced state and appears a dark, purplish red color. The consumer has continually rejected attempts to market meat in this color state, even though it will return to the expected color state soon after the package is opened. Extensive use has been made, however, of the vacuum packaging technique to distribute fresh meat in larger or wholesale cuts.

The conditions required for packaging cured meat are quite different from those for fresh meat. The major factor is that cured meats are packaged in a barrier film, under vacuum, to provide a low oxygen environment. This is the ideal condition to maintain the nitrosylmyoglobin color while at the same time and in combination with the salt and residual nitrite present limit microbial growth so that a shelf-life of several weeks is obtained. The low oxygen level also retards the process of oxidative rancidity and the off-flavors associated with it. The packaging material should also be a barrier to moisture transfer. Most of these barrier-type packaging materials contain several layers or components, and may be either flexible or rigid. Transparency is important to provide a good view to the consumer.

A variation that has found application in the industry is that of cooking the product in a bag. If done properly, purge is minimized or eliminated and post-processing contamination is avoided. Deli style cooked hams, some poultry products, such as turkey breast, and chunked and formed roast beef are examples. The products are handled either as a cook and ship or cook and strip. In the latter case the bag is removed after preparation so that additional procedures can be done on the product and it is then repackaged for distribution.

A procedure that has found wide acceptance in some European countries, but has not been used much in the United States, is modified atmosphere packaging of retail fresh meat. The process consists of packaging with a gas mixture. The packages are quite attractive and often consist of a well, containing the meat, over which is sealed a clear film. The gas mixtures consist generally of oxygen and carbon dioxide and sometimes nitrogen. The oxygen serves to maintain the bright red, oxymyoglobin color and the carbon dioxide provides a microbiological inhibitory action. Proper hygiene and temperature control are critically important, and expected shelf-life is between six to ten days.

The product is high quality and attractive and offers special considerations in the distribution chain. Obviously, there is a cost involved. Some concern has been expressed about potential microbiological problems. Depending on the carbon dioxide content the aerobic spoilage organisms may be selectively inhibited. These are the ones causing spoilage and associated off-odors that are a warning to consumers that something is wrong. Growth of the spoilage organisms provides a degree of competition (therefore inhibition) to pathogenic organisms should they be present. Considering this, and the extended shelf-life, the possibility exists for growth of pathogens.

Packaging is an enormous area of effort for both scientists and marketing specialists. It allows variation in processing and product development and is crucial in preservation or product stability. It is used for product identification, distribution and presentation.

CHAPTER 6

CHEMICAL METHODS OF PRESERVATION

Consumers have a strong preference and appetite for fresh foods. However, if they are informed that the fresh-appearing food has been preserved with chemicals, their zeal diminishes quite rapidly, and, in fact, they may reject the food and seek a "natural" or "organic" alternative.

This negative reaction to additives or chemicals in foods has been especially vigorous the past quarter century. That is not to say, however, that preserving foods with chemicals had never been a problem previously. It had, and at the turn of the century great concern about using chemical preservatives for commercial gain and to deceive the consumer resulted in legislation to protect the consumer.

Preserving meat by salting (and then by the process of curing) is an extremely old and well-accepted procedure. Similarly, the use of vinegar (acetic acid) to pickle meat, and spices (some of which contain naturally occurring antioxidants) has been quite acceptable because of the longtime tradition of safe use. However, consumers really raised questions when attempts were made to preserve meat with such chemicals as formaldehyde, sodium benzoate and boric acid—or to revive a poor color with sulfur dioxide.

Chemical preservatives should not be used as a substitute for poor processing conditions or to disguise an already spoiled product. They may be quite beneficial in combination with other forms of preservation, such as refrigeration, to provide an especially good protection for a perishable food. A combination of preservation methods is often used to optimize stability and prevent loss of product quality, while maintaining freshness and nutrient value. Chemical preservatives are used to prevent or slow microbial spoilage or the growth of pathogens. They are also used to prevent or slow chemical processes such as oxidation, and to maintain, for example, desirable color. In addition, chemical preservatives often impart desirable palatability properties, such as the special flavor and texture of cured meat.

The control and regulation of use of chemical preservatives in food are complex and intricate. They vary from country to country, are different for different foods and are constantly being reviewed. In the United States, the Pure Food and Drug Act was passed in 1906, and the Federal Food, Drug and Cosmetic Act was passed in 1938. The purpose of these laws was to prevent anything harmful to humans from being added to food. In 1958, the Food Additives Amendment was passed and prohibited the addition of a new additive to food without authorization. In addition, the additive had to be shown safe before permitted for use. Some additives were permitted based on prior sanction or the status known as GRAS (generally recognized as safe). The Delaney Clause prohibited absolutely any compound found carcinogenic to animals or man to be added to food. It is a complicated series of laws with interpretation made more difficult because of definitions and deciding who has authority (i.e., which federal agency, or federal versus state authority).

For meat, the major use of chemical preservatives will be covered by a discussion of the well-known curing process. Several other sections will deal with other components that can be considered as preservatives. Some compounds, such as carbon dioxide, are discussed elsewhere in the text.

CURING

Ideas about curing are presented first in this chapter about chemical preservatives because curing has had such a long history of use, actually combines several chemicals into one process and continues to be used extensively and successfully today. Not only is a definite preservative function imparted by curing meat, but also, an entirely new class of meat product is created; cured products have different appearance and palatability characteristics from fresh meat.

Salting was a means used very early in history to preserve meat. Curing evolved, as it was observed that some sources of salt caused a color change in the meat that eventually became recognized and desired. We know now that the color change was due to a nitrate impurity in the salt. It became known that saltpeter (potassium nitrate) was the critical ingredient for curing. How the color change was actually caused chemically remained a mystery until the final few years of the 19th century. Then, German scientists discovered that under typical circumstances the nitrate was reduced to nitrite (undoubtedly due to microbial action), and it was the nitrite, not the nitrate, which caused the color change.

At the turn of the century, work by an English scientist provided the basis for our understanding of the color change in cured meat (see Chapter 2 for diagram of reactions and compounds formed). The addition of nitrite initially causes an oxidation of myoglobin so that the brown color of metmyoglobin is evident. Due to reductants present in the meat, or added to it, the metmyoglobin is reduced again to the myoglobin form. Nitrite, or more correctly the nitric oxide formed from it, reacts with the heme portion of the myoglobin to form nitrosylmyoglobin that is a red color. When the meat is heated, the pigment is converted to dinitrosylhemochrome, which is the pink color characteristic of cured meat.

In the United States, empirical work by USDA scientists was undertaken and formed the basis of regulations issued in 1925 that today still control the use of curing ingredients. The 1906 Federal Meat Inspection Act permitted the use of sodium nitrate for curing. Because an amount of nitrate was not specified and due to the great uncertainty of conversion of nitrate to nitrite, many problems were experienced with the curing process. If sufficient conversion to nitrite did not occur then the curing process would fail and good color would not develop. If excess nitrate was added and good conversion to nitrite occurred there could be too much nitrite present that would attack the pigment and produce a green color. In essence, the regulations of 1925 permitted the substitution of sodium nitrite for sodium nitrate and controlled the level by stating that no more than 200 ppm of sodium nitrite could be present in the finished product.

Since the mid-1920's, many advancements have been made in the meat curing industry. The direct use of nitrite greatly reduced the time for curing, since waiting for reduction of nitrate to nitrite was no longer required. Moreover, the process became much more reliable and uniform because the level of nitrite could be much more closely controlled, compared to depending on an unpredictable conversion of nitrate to nitrite.

Engineering advancements, such as automatic, multi-needle injection equipment speeded up the curing process by allowing a more uniform distribution of the curing solution in large, intact pieces of meat.

The most significant improvement in the curing formulation was the introduction of use of reductants such as ascorbate or erythorbate. These greatly speeded the reduction of metmyoglobin that was formed from initial contact with nitrite. The result again was more uniformity because the biological reducing system of the meat was no longer depended on solely, and the quick and uniform reduction allowed the process to be automated instead of batch type. Residual reductant, by acting as an antioxidant, also maintained color better in the finished product. These reductants also participate in the chemical reactions by which nitric oxide is produced via nitrous acid from the nitrite.

The use of vacuum packaging for cured meat products greatly extended shelf-life, permitting a lengthy and complicated distribution chain that delivered high quality products to the consumer. Proper refrigeration is certainly a key component. Reduction of salt level used in curing has taken place due to desire of consumers to have a lower intake of sodium. Considerable research effort was undertaken to ensure that reducing salt level did not compromise safety and quality of cured products.

Finally, cured meats withstood a severe challenge during the 1970's. It was felt that they might be harmful to human health due to the possible presence of nitrosamines. A fuller discussion of this will be given at the end of this section on curing.

The ingredients commonly used to cure meat are salt, nitrite (nitrate), sugar, reductants, spices or seasonings and phosphates. The blend of ingredients may be rubbed on the surface of the meat as a dry powder. This is a specialized and oldtime procedure that requires extended periods for penetration of the chemicals and is used with such products as dry-cured or country style hams. Alternatively, the ingredients may be dissolved in water to form a curing solution or pickle. The meat pieces may be soaked in the pickle, or it may be injected into the piece of meat. Automatic injection is now used frequently for such items as hams and bacon and may be followed with a tumbling exposure to speed final distribution of cure and ensure it is thorough and uniform. Alternatively, the curing ingredients may be added during grinding or comminution as with products such as frankfurters or luncheon meats.

After incorporation and distribution of cure, the product is exposed to some degree of heating, and very often to a smoking process.

The results of curing are characterized by specific and characteristic color, flavor, texture and preservative properties. The color is a light pink, which is relatively heat stable but may fade on exposure to light. The specific flavor depends on the seasoning, but nitrite may contribute by itself to flavor—this may be related to the fact that nitrite slows oxidative rancidity and therefore the development of "off" flavors. The curing ingredients, especially salt, contribute to a general preservative effect, and nitrite has specific attributes in this regard —notably the inhibition of outgrowth of spores of *Clostridium botulinum* should they be present and should the product be abused.

Salt

Salt, or sodium chloride, is the major ingredient used in curing and generally is present in the finished product at a level of about 2.5%. Some years ago, cured meats had a higher concentration of salt and it was the main means of preservation. The level of salt addition was reduced as alternative and supplemental forms of preservation, such as refrigeration, came into use. The level has been reduced even further more recently as consumers have attempted to lower their intake of sodium for health related reasons.

It must not be overlooked that sodium is required for proper physiological functioning of the human and must be provided as a component of the diet. Moreover, salt is an effective inhibitor of microorganisms, imparts flavor, and in some cured meats, such as emulsion type products, serves an important functional role by solubilizing the myofibrillar proteins. Different organisms have specific, varying tolerances to salt, and in general, salt decreases the availability of water.

The purity of salt used in curing is important. Possible contaminants, such as metal ions, may have an undesirable effect, for example, in promoting oxidative rancidity.

Nitrite

Sodium nitrite is the vital ingredient for meat curing. Nitrate may be used with the aim of producing nitrite from it by microbiological reduction. It was thought some time ago that by having a reservoir of nitrate, then nitrite could be produced continuously from it and be present, more or less continuously, at a certain level instead of decreasing as it was used. While nitrate is still allowed for use today in some circumstances (such as dry cured products), almost all curing is done directly and solely with nitrite.

It seems a relatively simple matter that nitrite reacts with myoglobin to produce the color characteristic of cured meat. In fact, nitrite is an extremely reactive chemical. When it is added to a complex biological system such as meat, the result is a very complicated situation. Nitrite is soluble in water, and at the mild acid conditions of meat only small quantities exist as nitrous acid. Further reactions produce nitric oxide that is probably the active nitrosating agent. It is known that nitrite participates in nitrosation, diazotization and deamination reactions. With meat, nitrosation is the most important with the possibility of nitrosation reactions occurring at nitrogen, carbon and sulfur atoms.

When nitrite is added to meat in the usual curing procedure about 50% of it disappears quickly from analytical detection. Depending on the type of product, extent of heat processing and other factors, the so-called residual nitrite continues to decline, reaching a level of perhaps only 10% or less of the original amount added. Only about 5 to 15% of the added nitrite reacts with the myoglobin. A small quantity reacts with the lipid component and most of the lost nitrite reacts with other non-heme proteins. It remains a question if such nitrite is indeed chemically reacted and firmly bound or in fact if it is more loosely bound and could participate subsequently in transnitrosation reactions. A potential major problem is the possibility that nitrite reacts with secondary amines to form nitrosamines, which as a class of compounds are known for their carcinogenic properties.

The antimicrobial activity of nitrite is a well-known property. It was established already in the early 1950's that nitrite afforded specific protection against outgrowth of spores of *Clostridium botulinum*. This was especially important as cured meats and canned hams were being put in vacuum packages. Both types of cured meats had not had sufficient heat processing to inactivate spores, and in both cases the packaging provided anaerobic conditions required for outgrowth of spores if temperature abuse occurred. "Temperature abuse" is used to mean that the product is abused by exposing it to elevated temperature rather than keeping it at recommended refrigeration temperature.

During the 1970's several large inoculated pack experiments were conducted cooperatively by the industry and government. They proved beyond doubt that nitrite by itself did have a specific protective effect. Although nitrite was thus singled out, it was also emphasized that the interactions with salt level, pH and thermal process were all important in the total picture. From a mechanistic viewpoint it is possible that nitrite may act by blocking sulfhydryl groups or by reacting with iron-containing compounds.

Sugar

Sugar is normally included in curing formulations for a flavor function and to moderate the potential harsh flavor associated with high salt concentrations. Obviously, the latter becomes less important as the amount of salt being used is decreased. The type of sugar used may influence color development (that is darkening or browning) as the meat is exposed to heat during preparation. In the case of cured, fermented products, the sugar is necessary as an energy source for the organisms performing the fermentation. With the amounts of sugar used in curing, there is little if any preservative effect from it.

Reductants

Ascorbic acid, erythorbic acid (isoascorbic) and their salts are commonly used in modern curing practice. They are used primarily to speed up the curing process and to make it more uniform. Ascorbate, for example, is active and effective in reducing metmyoglobin to myoglobin. It also promotes the formation of nitric oxide from nitrous acid. Finally, the residual reductant remaining acts as an antioxidant prolonging shelf-life.

Spices and Seasonings

These substances are added to give a characteristic flavor. They are not generally recognized as providing a preservative effect, with the possible exception that some may contain naturally occurring antioxidants.

Phosphates

The alkaline polyphosphates are used in the meat processing industry with sodium tripolyphosphate being the most common. Sodium acid pyrophosphate is known to speed the development of cured color, but it also has detrimental effects on water binding and emulsion stability. The phosphates have been used most often in the United States in production of ham and bacon, but now are also allowed for use in cooked sausages. Some problem may be encountered in their use due to low solubility, sometimes, and the possibility that they may impart a metallic or soapy taste to the product.

The major purpose for their use is to increase water-binding and improve yield. The mode of action is primarily by increasing pH that in turn increases water-binding. Other possibilities for mode of action include increasing protein hydration by making more sites on the proteins available due to alteration of structure or by binding certain ions that might otherwise interfere with water-binding. Phosphates may produce a preservative effect in that they retard development of oxidative rancidity and may also slow microbial growth. These actions probably result from the fact that they bind heavy metal ions.

Regulatory Control

Regulatory control of meat curing has not been explained because it varies from country to country and by product and is changing as new information is produced. The regulatory process is generally to limit use of nitrite to that needed to produce an effect and to minimize any possibility that any of the substances used may be harmful to human health.

This latter point is illustrated by the severe challenge directed during the 1970's at the continued use of nitrite as a meat curing agent. The major concern was that nitrosamines could be formed in meat due to the nitrite curing process. Nitrosamines are recognized carcinogens. Of secondary concern was the fact that cured meat contains residual nitrite that might nitrosate secondary amines in the acid gastrointestinal environment and form carcinogenic nitrosamines. If nitrite was banned as a curing agent then the protection against botulism would be lost. The entire episode is too lengthy to relate here but suffice it to say that the challenge was judged unfounded as being of real concern to human health. In the process, however, the curing industry did make several improvements. Only the minimum amount of nitrite was used for a given application. The use of nitrate was largely eliminated. Maximum use of reducing agents was made, since they had been demonstrated to inhibit formation of nitrosamines.

SMOKE

Smoking is almost an integral part of curing, as is heat processing. However, it is discussed separately here because the process generates some chemicals with preservative properties that are deposited on the surface of the product being smoked. This is in addition to the preservative effect associated with the heat possibly generated during smoking and from the physical effect of dehydration usually occurring at the surface.

A "skin" or compacted layer is formed at the surface. So the smoking process is rather effective in forming a type of physical, surface barrier.

Moreover, the smoking process lowers microbiological population due to heat and the deposits of chemicals that have bacteriostatic action. These circumstances may be altered if a liquid type of smoke (compared to natural vapor smoke) is employed—and it also depends how it is applied.

Smoke for meat is normally generated by a slow combustion of hardwood sawdust. Originally, this was done in simplest form merely by hanging meat in a room that contained a smoldering fire. Much progress has been made in design and engineering of equipment so that the standard now is the so-called air-conditioned smokehouse, which besides generating smoke has provision for accurate control of temperature, humidity and air flow. The same concept has been adopted to continuous manufacturing equipment in which the product is moved through zones of smoke and heat. Smoke generators have also been refined to give more control over operation at suggested temperature and air availability.

Smoking is done for flavor and appearance as well as for the preservative and antioxidant effects. Color development, besides the pink color associated with nitrite curing, is primarily a brownish or mahogany. The smoke color is a surface phenomenon of the Maillard type in which carbonyls in the smoke react with free amino groups of the meat proteins. The desired degree of smoke for any given product is achieved primarily by balancing smoke density, air velocity and humidity (moisture at the product surface) within a given period.

Smoke is a complex mixture containing over 200 identified compounds. The composition of the smoke depends not only on the wood source but also on the important factors of temperature of combustion and available oxygen during combustion. Smoke contains two phases—a gas or vapor phase and a disperse or particulate phase. The latter is not of much consequence since little of it is actually deposited on the surface of the meat. On the other hand it is the components of the gas phase that are deposited on the surface and interact there with the meat components.

The classes of chemical compounds found in the vapor phase are acids, phenols, carbonyls, alcohols and polycyclic hydrocarbons. Examples of organic acids are formic, acetic, propionic, butyric and isobutyric. Larger molecular weight acids would be found in the particulate phase. The major function performed by these acids is to coagulate or denature the meat proteins at the surface of the product. This is important if the casing is to be removed, as in manufacture of weiners where the cellulose casing is stripped. Without the coagulated protein layer, peeling of casings can be extremely difficult.

The phenols play an important role, with perhaps the most beneficial being an antioxidant function. They also give rise to a smoky flavor and provide a degree of bacteriostatic action.

Alcohols form a class of compounds that functions primarily as carriers for other chemicals. Methanol is the most common alcohol found in wood smoke.

There are numerous carbonyls present. The short chain, simple carbonyls are important in imparting a smoke flavor and in giving a desirable color—especially an attractive brownish cast.

Many different polycyclic hydrocarbons are present with probably the most well-known being benz(a)pyrene—because it is a carcinogen. The hydrocarbons are not known to impart a flavor or preservative effect.

During recent years, considerable interest has been generated in the use of liquid smoke preparations as contrasted to the natural vapor smoke. In fact, they are used extensively in the industry today and in the future will probably replace in large part the use of natural smoke. There are several reasons for this. Liquid smoke preparations allow a more constant, uniform and repeatable production of product. Expensive equipment such as smoke generators are not required and the process can probably be moved along more rapidly. The environmental problem of how to deal with emission of smoke into the air does not have to be solved. And finally, the liquid smoke preparations have had carcinogenic compounds removed during their preparation.

Application of liquid smoke to meat products may be accomplished in several different manners. It may be mixed or incorporated directly into the product. More likely is the use of procedures incorporating spraying or dipping, or alternatively the liquid smoke can be atomized or vaporized into the smokehouse or cooking chamber.

One method of preparation of liquid smoke is to smolder the sawdust under controlled oxidation conditions, and then draw the smoke generated countercurrent to water through an absorption tower. When the preparation reaches a given concentration, it is aged, filtered and then ready for use.

ANTIOXIDANTS

Numerous considerations and regulations impinge on the use of the preservatives known as antioxidants. The basic goal in using such preservatives is to slow or prevent oxidation and the resultant development of rancidity, off-flavors and discoloration. The situation is further complicated due to the interrelationships of autoxidation and microbial spoilage, both of which are occurring simultaneously. In the total picture of meat preservation, microbial spoilage is probably more important, although oxidative changes can result in consumer rejection of the product.

There are naturally occurring antioxidants. For example, some spices, such as rosemary, which can be used as seasonings in meat, have antioxidant activity. However, most antioxidants considered here are viewed simply as chemical additives by the consumer. While the consumer demand for freshness is strong, so is the desire for freedom from additives.

To prevent or minimize oxidative changes, oxygen concentration can be lowered—for example, by vacuum packaging. Another possibility is to use an enzyme reaction to remove oxygen. Many antioxidants function by scavenging free radicals and interrupting the chain reaction or propagation. Chelators may be used, primarily to bind metals such as copper and iron, which are known catalysts for the reactions that form free radicals. Finally, oxidative enzymes may be inactivated.

A combination of ingredients and procedures may be the best choice to prevent oxidative changes. The efficacy of the system, regulatory matters, ability to survive processing steps and avoidance of imparting palatability problem are all important. Moreover, a synergy may be attained with combinations of two or more antioxidants. Incorporation and distribution of antioxidants into meat must consider solubility in water, solubility in lipid and solubility at the lipid-water interface.

The two major classes of meat—that is, fresh and processed—are treated differently. The former is extremely perishable, but it is displayed in the presence of oxygen to maximize oxymyoglobin formation. Regulatory control generally dictates that additives are not allowed. Processed meats are quite a different matter. Various chemicals are permitted, the finished product may be quite well-adapted to vacuum or sealed storage, and some extent of heat processing is usually employed during manufacture.

There has been a special application in fresh retail meat—the so-called Wilson's TenderCuts. It is actually an attempt to move toward branded, centrally prepackaged fresh pork. The process is patented, and, besides the chemical additives involves special cutting, chilling and packaging equipment and conditions. The process is allowed by a USDA regulation permitting addition of several GRAS substances to help maintain color. The regulation was issued on the basis of the procedure being safe, wholesome and not masking spoilage. Use of the following chemicals is permitted: ascorbic acid, erythorbic acid, citric acid, sodium ascorbate and sodium citrate. The package must be labelled with a qualifying phrase contiguous to the product name indicating the meat has been treated to maintain color. The Wilson process also involves pumping the loins with a solution of phosphates, dextrose, citric acid and salt to keep the meat from cooking up dry and tough. The surface of the cut meat is sprayed with the other additives, and the retail-ready package has a high CO_2 atmosphere. The packages have a minimum 14-day shelf-life and the product is sold for a premium price.

Some antioxidants are considered oxygen scavengers. These substances are also reducing agents that act by transferring hydrogen atoms. The most commonly known are ascorbic acid (vitamin C) and erythorbic acid (D-isoascorbic acid), which are also available and used in the sodium salt form. As discussed previously, they are used widely in cured meat to accelerate the curing process, and they also act subsequently to stabilize both color and flavor. Sodium ascorbate is used specifically in bacon to lower the residual nitrite and inhibit formation of nitrosamines. With exception of the Wilson process noted above, these substances are not permitted for use in fresh meat.

Ascorbic acid is soluble in water. It removes oxygen and in the process is converted to dehydroascorbic acid. It may be most effective, sometimes, by regenerating other antioxidants through donating hydrogen atoms. It may also contribute a high synergistic effect, for example, in combination with vitamin E. It is GRAS.

Erythorbic acid is also GRAS and is also a strong reducing agent.

The issue is somewhat confused because the two above substances have both a pro-oxidant effect and an antioxidant effect depending on the circumstances. The result is determined by such factors as concentration, presence of trace metal ions and pH. Several experiments have appeared in the literature and describe various attempts to use these substances in fresh meat to increase shelf-life by spraying, dipping or otherwise incorporating them. With proper conditions they do delay formation of metmyoglobin.

Sulfites have both antioxidant and antimicrobial activity. Their use is not allowed in meat in the United States, but a brief discussion will follow in the subsequent section.

Another possible method to remove oxygen is by use of the enzyme system glucose oxidase. It catalyzes reaction of oxygen with glucose to form D-gluconic acid and hydrogen peroxide. Catalase is a constituent of commercial glucose oxidase systems, and it prevents accumulation of hydrogen peroxide by hydrolyzing it.

Some antioxidants act by ending the free radical chain reaction. These substances act by contributing hydrogen ions being converted themselves to free radicals that are stable and do not propagate further oxidation of lipids. Commonly recognized antioxidants in this class are butylated hydroxyanisole, butylated hydroxytoluene and tertiary butylatedhydroquinone. These are all highly soluble in lipid and find their greatest use in stabilizing oils.

Propyl gallate is slightly water soluble and sensitive to heat. Since it binds iron atoms and produces a dark color, it is used in combination with citric acid as a chelator. It finds application in meat, being used commonly in fresh pork sausage.

The tocopherols find more use in animal fats than in vegetable or plant oils. They are naturally occurring, usually in plants. Alpha tocopherol is the most active form of vitamin E. Tocopherols are miscible with fats and are heat labile. There has been recent strong interest in their use in fresh meat by incorporating them via dietary feeding of the animal. In beef for example, feeding vitamin E to the animal has been shown to greatly extend color shelf-life of steaks.

Chelating agents are not antioxidants as such but rather function by complexing pro-oxidant metals such as iron and copper. Commonly used ones are citric acid, polyphosphates and EDTA. Polyphosphates have been discussed previously regarding their use and functions in meat products. Citric acid does find use in meat, and is normally combined with an antioxidant, such as ascorbic acid, making an effective combination to prevent oxidative rancidity. It is used in fresh pork sausage, some dry sausages and dried meats.

SULFITE

Sulfur dioxide, sodium sulfite and sodium and potassium bisulfite and metabisulfite comprise the group of compounds known collectively as sulfites. They are employed as weak antioxidants and in some cases are strong antimicrobials. Through the years they have been employed as preservatives in various foods and beverages. They are effective, for example, to preserve the freshness and prevent discoloration of cut fruits and vegetables as might be offered in a salad bar. Recently, however, use has been removed from GRAS listing for fruits and vegetables to be served, sold or presented raw. Also, in foods containing 10 ppm sulfite, despite source, a label so stating must be used. The concern is for individuals who are sensitive or allergic to sulfites.

These compounds are not allowed for use in meat in the United States. They are recognized as relatively strong preservatives. Moreover, use of only a small amount in meat imparts a very bright red color. Sulfites are also known to degrade the vitamin thiamine, of which meat is a good source.

SORBATES

Sorbic acid (2,4-hexadienoic) and its salts are effective and rather widely used preservatives. In the case of meat in the United States, however, the use of sorbic acid is restricted to inhibiting mold growth on the surface of uncooked and dried sausages. This is normally accomplished by dipping or spraying the casings. In contrast, sorbic acid is widely used in Japan for preservation of meat products.

Experimentation has shown that sorbic acid is an effective inhibitor of mold and yeast growth as well as an effective antimicrobial agent in meat. It has also been studied as a nitrite sparing agent in cured meat because of its antibotulinal activity. However, concern (based on sensory testing of sorbate-treated bacon) that it may be harmful to human health has kept it from being approved for use.

LACTATE

There has been strong interest in the potential use of sodium lactate as a preservative for meat. The idea is attractive because lactic acid is a natural constituent of meat, and sodium lactate has been approved for use in both mammalian meat and poultry. It has been shown to have antimicrobiological activity against a broad range of organisms. Interest is especially evident in using it to control pathogenic organisms, such as *Clostridium botulinum*, in products where there are not many lines of preservation defense in the event of product abuse. The suggested mechanisms of action include feedback inhibition, intracellular acidulation, interference with proton transfer across cell membranes and altering water activity.

The preservative effect has been demonstrated in both fresh and cured meat. The amount of sodium lactate added ranges up to 7%, with it more usually being 4% or less. Inhibition is increased as higher levels of sodium lactate are added. Above a certain level, palatability may be influenced.

ACIDULANTS

Acidulation or the addition of acid to lower the pH of a food is a historical method of preservation. With meat, the most common way to acidulate is to make a fermented product so that the meat more or less acidulates itself (discussed in detail in the Chapter 7).

Direct acidification has been used to make, for example, pickled types of meat products. Vinegar contains about 4% acetic acid and is used to make pickled pigs feet. Jellied products such as sulze have vinegar added directly. The acid has a sour flavor that can be quite sharp, and having the acid concentration high enough to be a strong preservative by itself is difficult without making the product too sour for consumption.

The undissociated form of the acid is responsible for the antimicrobial properties. This may be explained in that the nonionized acid can cross the bacterial cell membrane, and once inside the cell become ionized again. pK_a is the pH at which 50% of the acid is undissociated, and for most organic acids this is in the range of pH three to five. Therefore, in meat, as the pH is lowered the concentration of undissociated form increases and a greater preservative effect is obtained. At a higher pH of 5.5 to 5.8 most organic acids are relatively ineffective. The concentration used is limited by flavor aspects as well as influences, for example, on color. Organic acids are effective until removed from action by volatization, metabolism or binding.

With direct addition of acid into meat products there are several potential problems. Basically, it is a matter of too much too fast. In most meat products the acid must be added before heat processing. That means that the acid can act on the native meat proteins before they have been heat denatured. The result may be a change in texture or "set-up" of the product. In emulsion type products the result may be a release of the fat causing it to leak from the product. Flavor may be a problem—limiting the amount that can be used and not giving the proper flavor balance. For example, the direct addition of lactic acid may give quite a different flavor compared to a lactic acid producing fermentation. In the latter case there is a slow production, during fermentation, of acid and other metabolites that contribute to flavor.

Glucono-delta-lactone has been used in meat products to avoid the potential problems of direct addition of an acid. This compound is hydrolyzed in the meat system as it is exposed to water and as temperature is increased during heat processing. As a result, the pH is decreased as the amount of gluconic acid increases. There is a limit as to how much pH decrease can actually be obtained due to regulation on the amount of Glucono-delta-lactone that can be used.

Another procedure for direct acidulation is to encapsulate the acid to be used. This is normally done in a fatty substance that releases the acid as heating occurs.

CARCASS RINSES

Considerable research has been undertaken recently to determine the value of using organic acid washes or rinses of hot carcasses as a means to control microbiological contamination. Lactic acid or acetic acid in the concentration range of 1 to 2% are effective without causing color or flavor problems. Action is via a pH drop and by the effect of the undissociated form of the acid. It has been shown that a rinse with trisodium phosphate can effectively reduce the incidence of *Salmonella* on raw chicken. Obviously, the key is to use such procedures as part of a total hygiene program, and, of course, governmental approval is a consideration.

CHAPTER 7

MICROBIOLOGICAL METHODS OF PRESERVATION

This chapter focuses on utilization of the microorganisms residing in and on the meat, or intentionally added to it, rather than on extrinsic control factors. So, the major considerations are number and type of organisms present and their interactions with each other in the environment of the meat.

Fresh meat and cured meat represent two very different environments. In fresh meat, initial contamination comes mostly from the animal and the people slaughtering it with the equipment used. The nutrient environment is quite ideal for microbial growth, and the most easily employed barrier to growth of microorganisms is reduced temperature. With processed meat, the population of organisms present normally results from recontamination following heat processing. The organisms find themselves in a more hostile environment usually including, for example, salt and nitrite and often a packaging that results in anaerobic conditions.

Ground meat presents a special situation since it is still fresh but has been exposed to further microbiological contamination during the particle reduction and mixing associated with grinding.

The microbial situation is quite dynamic, with several species often present. They must compete among themselves in the biological quest to grow and reproduce. The dominant organisms will increase in numbers and the weaker will decline. Various organisms may be synergistic or antagonistic to growth of others. The substrate environment is also dynamic as nutrient concentrations change and the metabolites present are altered. The result may be waves of different organisms predominating and subsequently declining; this process is known as microbial succession.

If the environment is altered, microorganisms respond with homeostasis mechanisms to keep their physiological systems in balance. This places increased energy demands on the cells and serves to help in preservation attempts.

Fermentation is a time-tested method of using specific organisms in a preservative role. In the past quarter century, the fermentation process has been highly refined—not only to be more effective in production of acid but also to be more controllable, resulting in the production of a very uniform product with specific palatability traits.

Overall, the idea of using microorganisms themselves to accomplish a degree of preservation holds considerable promise for the future. The competition within a meat system, in which the spoilage organisms usually win out over the pathogens, is an extremely important protective mechanism for the consumer.

Product is usually spoiled or organically unacceptable before pathogen populations have time to increase to potentially hazardous levels.

It has been known for some time that microorganisms produce antibiotics or similar compounds that in fact have preservative effects. These are known generally as biopreservatives, and current research is directed at explaining more about the preservative mechanism. New molecular techniques may allow the design of organisms capable of producing specific substances.

The possibility of using the microorganisms present naturally in the meat to exert a degree of preservation is quite acceptable to the consumer—especially in contrast to adding a chemical preservative.

COMPETITION

Competition for survival among the mixed population of microorganisms normally present in meat can be strong and specific. As a matter of fact, this natural competition or selection is an effective barrier or hurdle in the total picture of preservation. In cooperation or conjunction with other preservation factors such as cooling or salting, a narrow and specific pathway can be set for predicting the outcome. Similarly, an unknowing or not knowledgeable selection of the total set of conditions for meat may open the way for growth of an undesirable organism. There are essentially two factors to consider—the environment and the population of organisms.

The environment consists of the nutrient base and the external factors applied. The nutrient composition of meat provides an excellent medium for growth of microorganisms. Proteins, vitamins and minerals are readily available, as is an abundance of water. Packaging conditions dictate in large part the level of oxygen available.

External factors can be thought of as stressors. These are the classical preservation methods such as heat, cold and salt, which have been discussed at length in previous sections.

Within meat, there are in fact microenvironments. Most notably (in non-vacuum packaged products) there are different environments at the surface, near the surface and in the deep interior of meat. Clearly, for example, oxygen cannot penetrate completely into the depth of meat. In intact pieces of meat there are different compositional areas such as islands (or areas) of lipid. Additionally, the connective tissue framework and elongated tubular, cellular structure of muscle provide not only barriers but also channels for movement of substances and, occasionally, microorganisms.

Finally, this environment is altered as microbial growth occurs and other postmortem changes proceed. Not only do chemical factors such as pH and nutrient availability change, but also products from microbial metabolism accumulate. Also, the structure undergoes some degradation with, for example, cell membranes becoming more permeable.

The other factor of importance is the population of microorganisms—not only the kinds but also the number, both of which depend on the initial contamination. Besides growth, which has been discussed previously, the matter of virulence of pathogenic microbes is important. In general, the expression of virulence is more sensitive to aspects of competition and environment than is mere growth. As an example, *Staphylococcus aureus* may grow but be incapable of forming its heat-stable enterotoxin.

It has been shown that in fermented products the microorganisms present are not evenly distributed throughout the matrix but are clustered in cavities or "nests". This implies the possibility of concentration and diffusion effects.

An example of using microorganisms in a competitive situation to ensure food safety is the so-called "Wisconsin process for bacon". A culture of lactic acid bacteria and a fermentable carbohydrate are added to the normal package of bacon. If the package is subsequently temperature abused, the lactic acid bacteria will grow, lower the pH and compete with other organisms present, which will provide a preservative effect.

FERMENTATION

Fermentation of meat products is an ancient technology. It has the dual purpose of providing a strong preservative action while simultaneously creating specific products with rich flavors that very often command premium prices.

Through the years, the fermentation process in meat was a sort of art form—the result could be either a very special product or a failure. The procedure depended on using the naturally occurring flora and handling the meat in such a way (i.e., time, temperature) as to give the process the best chance for success. Some advancement was made by back-inoculating, or taking a portion of a batch of sausage and adding it into a new batch so as to ensure that the proper microorganisms were present.

In the mid-1950's, work was announced that encouraged development and use of starter cultures in meat fermentation. In reality, separate and independent work led to the same conclusion that subsequently allowed the industry to make enormous technological advances. Niinivaara of Finland worked with *Micrococcus*, and Deibel and Niven of the United States used *Pediococcus*, and both put forward the same idea that starter cultures could indeed be used.

The usual pH range of meat is 5.7 to 6.2, and typical fermentation may bring it down to as low as pH 5.3 with some processes going down to pH 4.6. Most fermented products lose some amount of water and many undergo a substantial drying—a process that proceeds well at the lower pH reached during fermentation. The isoelectric point of the muscle proteins is between pH 5.1 to

5.3, and is the point at which water is most loosely bound and therefore more easily lost from muscle. Low pH is not the only preservative factor in fermented meat but functions in combination with lowered water content and the presence of salt and nitrite (possibly nitrate). The total amalgam of conditions attained give a strong general preservative action against spoilage organisms and most pathogens.

There are two fundamental types of fermented meat products—dry and semi-dry. The dry is also a low-acid product with the pH normally not going lower than pH 5.5. It is important in this case that the fermentation be conducted at a low temperature, and proper drying is critical. It is a long-term procedure often requiring up to 60 days for completion. Most pathogens are inhibited, but *Listeria* and staphylococci do have a chance to survive the process. These products have characteristically strong flavors and odors resulting from degradation of lipid and proteins during the prolonged fermentation and drying phases. Salami is a typical example. The semi-dry product is high-acid by comparison, with the pH being 5.3 or possibly lower. A carbohydrate source is usually added so that the fermentation can proceed to the lower pH level. By present day practices, a starter culture is usually used. A typical example is summer sausage.

Dry sausages have a moisture to protein ratio of 2.3:1 or lower and water content between 25 to 45%. Semi-dry sausages have a moisture to protein ratio higher than 2.3:1 and a water content of 40 to 50%.

The earliest, uncontrolled fermentations required the completion of two processes. In addition to the fermentation that resulted in the lowered pH, it was necessary to reduce the nitrate, which was added for curing, to nitrite. micrococci and coagulase-negative staphylococci not only have the nitrate reducing ability but also are tolerant of salt and nitrite. However, they are sensitive to acid. So, it was common to hold the sausage mixture at a low temperature to permit nitrate reduction and then increase temperature so that other organisms could ferment the added carbohydrate, lowering pH. Now it is common to add nitrite directly and bypass completely the nitrate reduction step.

ANTIMICROBIALS THAT OCCUR NATURALLY

Many compounds occurring in natural products have displayed inhibitory action against microorganisms. The challenge is to assess if the natural product could be used in some way, for example, in a meat product, or if the active substance could be isolated, purified or concentrated and then incorporated into a meat system. Obviously, there are several questions to answer in attempting to use natural products. Do they effect changes in the normal characteristics of the meat? What is the extent and specificity of inhibition? What are the possible regulatory considerations? How stable are they in meat?

Lysozyme, which is a component of egg white, is a known antimicrobial that acts by attacking the bacterial cell wall. While gram-negative organisms are usually resistant, it is active against gram-positive pathogens such as *Clostridium botulinum* and *Listeria monocytogenes*.

The preservative effect of organic acids has been discussed previously. They also occur widely in nature, and, for example, benzoic acid is a component of cranberries. Other known inhibitors are plant essential oils such as those occurring in garlic and onion. Caffeine has antimycotic and antibacterial action.

The lactoperoxidase antimicrobial system has been investigated in milk. The system consists of three components—peroxidase, thiocyanate ion and hydrogen peroxide. The peroxidase and thiocyanate occur naturally in a wide variety of animal tissues. Hydrogen peroxide is generated readily by catalase-negative lactobacilli. Lactoperoxidase reacts with hydrogen peroxide, which in turn oxidizes the thiocyanate to hypothiocyanate or other substances, which are known antimicrobials.

Potential problems in attempting to activate this system in meat include compartmentalization within the cellular and connective tissue structure, potential effect on color of meat and the possibility that endogenous (or added) reducing systems in meat may reduce the hypothiocyanate before it can exert its antimicrobial action.

Hydrogen peroxide is by itself an antimicrobial. It may accumulate if the catalase enzyme is not present.

Diacetyl is another potential metabolic product from certain of the lactic acid bacteria. It has a strong buttery aroma and is required in high concentrations for a preservative effect.

Since meat derives from the live animal, it is appropriate to comment about the natural control of microorganisms in the living animal, before it is converted to meat. The most effective and naturally occurring antimicrobial system is the mammalian phagocytic cell. Not only do these cells migrate to the site of trouble by some chemotaxic response, but they are also extremely effective in damaging prokaryotic cells without harming eukaryotic host cells. The cell membrane of the organism to be attacked is obviously of key importance. Preservatives may damage it and alter the physiological function of the cell or they may change permeability of the cell membrane so that an agent can pass through and then act within the cell.

BACTERIOCINS

Bacteriocin has been used as a term for many groups of colicin-like antibiotics. An antibiotic is a substance produced by a living organism that is inhibitory to or destroys another living microorganism. Bacteriocins differ from other antibiotics principally due to the relatively restricted range of activity they have regarding other strains and species. Each bacteriocin is produced by a strain of bacteria, and in general is active only on other strains of the same or closely related species. Many different groups of bacteriocins are known. Colicin is an antibacterial substance discovered more than 60 years ago that is produced by a strain of *E. coli*.

Nisin is a microbiologically-derived preservative that has found use in food preservation. It is a peptide antibiotic produced by *Lactococcus lactis*, and it has been known for about 30 years. It inhibits gram-positive organisms but has no effect against yeasts or molds. Of special interest is the fact that it prevents outgrowth of germinating bacterial spores. Reports have shown variable success from using nisin in various meat products. While it has been used in a number of other countries, it has not been allowed for food use in the United States until very recently, when it was approved for use in certain pasteurized cheese spreads to inhibit outgrowth of *Clostridium botulinum* spores. This may signal that more microbiologically produced substances will be considered for food use in the near future.

As discussed previously, lactic acid bacteria are used successfully as cultures in fermented meat products to produce specific palatability and textural properties. They also occur in vacuum packaged fresh meat and vacuum packaged cured meat, and there is increasing interest in the occurrence and role of lactic acid bacteria in modified atmosphere packages of fresh meat.

A known effect of the lactic acid bacteria is inhibition of various organisms such as psychrotrophic enterobacteriaceae and *Brochothrix thermospacta* as well as *Salmonella* and *Staphylococcus aureus*. The effect is due primarily to production of acid, and secondarily to competition for nutrients. Intense interest has been shown recently in the possible production of bactericidal proteins (bacteriocins). This does in fact occur and may be especially important to inhibition of pathogens such as *Listeria monocytogenes*. In the future there will most likely be increased use of lactic acid bacteria as protective cultures or biocontrol systems.

Lactobacillus reuterii produce a low molecular weight, non-proteinaceous substance called Reuterin. It exhibits broad spectrum inhibitory action against spoilage and pathogenic organisms in food.

The bacteriocins, or bacteriocin-like compounds produced by the lactic acid bacteria, have several distinguishing characteristics. They are proteinaceous and inactivated by proteases. Their antibacterial activity is narrow spectrum and usually targeted at strains within closely-related species. They are reasonably heat tolerant and active at relatively low pH. The fact that they are broken down by proteases means they are inactivated in the digestive system following consumption. Therefore, the possibility that they may continue functioning and develop resistant species is not a concern to human health.

GENETIC ENGINEERING

Opportunity for meaningful advancement in technology exists with the use of genetic engineering. It is quite foreseeable that the lactic acid bacteria could be gene engineered to express different quantities of a given protein. As discussed above, the bacteriocins may be quite important in this regard. Genetic engineering could play a significant role with specific flavors and stable color. Regulatory considerations and consumer concerns must however be overcome before the new technologies can be fully used in food situations.

While the usual preservation methods using control of temperature and moisture are quite effective, there is a significant energy cost. The possibility of using the bacteria themselves, as discussed in this chapter, is a worthwhile area to explore. The use of starter cultures has become quite broadly employed, and the level of sophistication has increased. The starter culture procedure may lend itself well to the new techniques of genetic engineering. Technically, it is much easier to alter the genetic material of a bacterium than of a higher animal. For example, it would be much more difficult to alter specific properties of the meat by changing genetics of the animal than it would be to change a microorganism, which could then be placed on the meat to do a specific function such as produce a preservative or maintain color.

CHAPTER 8

MANAGING PRESERVATION

We examine now how all the pieces can be made to fit together to obtain ideal preservation of meat for a given set of conditions. It is a matter of understanding and integrating all considerations to obtain the best result. The possibility exists to put into action all of the facts so that a total program can extract the greatest benefit.

Two key issues surface immediately. What is the system? Who is responsible?

With better preservation, the producer or processor is not only assured of a higher quality product and a safer one, but will obviously have a more successful business venture. The consumer will also be more satisfied with the best quality and safety to be offered.

There are three fundamental areas for action or control. These are: (1) the raw ingredient, (2) the total environment in which it is converted, and (3) handling of the final product. These conditions vary markedly for different phases of the meat industry—for example, slaughter, processing or converting and retailing. However, there are also great similarities. The starting material must be good if a quality end product is wanted. If ingredients and procedures are not high quality, then a high-quality raw material is rapidly downgraded with the result of a poorer final product. Hygienic standards for and level of personnel training is of significant consequence.

In slaughter, the raw material is the animal received. The critical aspects are quality, health and cleanliness. Any of these factors can influence quality of the meat that will be produced. For example, manure and dirt attached to the hair and skin of the animal can lead to subsequent heavy and undesirable contamination of the meat with microorganisms. Similarly, precautions must be taken to ensure that the animal is free of prohibited or undesirable chemicals. Of course, handling of the animal must be such as to afford minimum stress. The slaughter process is primarily a disassembly. Proper slaughter and inspection procedures, and hygiene of equipment and personnel are obviously important. In all instances, knowledge about and control of the water supply, and continued vigilance against rodents and insects are high priority. Chilling procedures can markedly affect final quality of the meat.

The situation with the processor is different. In former times many slaughterers were also processors, so they knew the history (therefore quality, composition and properties) of the meat they would use for further processing. Today, slaughtering is quite specialized and processors purchase meat from companies who do only slaughtering. It is then the responsibility of the processor to determine the properties of the raw meat. Composition, microbiological load and temperature history are a few of the items that may be monitored. While it is impossible to generalize completely, processing usually involves one or more of the following—size reduction, incorporation of other ingredients, and heating. Incorporation of other ingredients allows the possibility of adding unwanted compounds or microorganisms and so must be controlled carefully.

Heating lowers the population of microorganisms and brings the manufactured product to an almost pasteurized condition. Then, recontamination becomes the issue of concern. An important idea is the constant inactivation rate observed for most bacteria in a given time-temperature process. This means that a higher initial load results in higher remaining numbers for a given time-temperature treatment. Therefore, the importance of beginning quality is again reemphasized.

Packaging is an especially important area for processors, since the product is often shipped in the package in which it will be sold, and it may be intended to have a shelf-life of several weeks in that package.

Retailers normally deal with two items—fresh meat and processed items. In the case of the latter, presentation involves putting the premade package out for display where case temperature may well be the most important factor regarding preservation. Fresh meat is usually received as a primal or larger piece and is then cut to retail size, wrapped and displayed. So, the important considerations are quality and condition of the meat received and then the sanitary and hygienic conditions of equipment and personnel. Display temperature is again an important factor, and of course, lighting is also important in determining appearance.

Employee training and (or) certification is important—not only for the people managing the programs, but for the production line workers. In the previous case, the managers must be well-versed and capable, and must be kept updated. The production workers should be made aware of procedures and why they are being employed.

QUALITY CONTROL

Quality control is a process used to control or standardize a product. In the meat industry, it is commonly employed to monitor the meat and other ingredients entering the plant, to control the manufacturing process and to inspect the final product. Depending on the aim and intensity of the program devised for a certain situation, there may be a significant amount of inspection, sampling and testing. Obviously, this can be laborious and costly. Examples of typical testing procedures are analytical composition (i.e., fat, protein and moisture), total bacteriological plate count and package weight.

The program is a management tool by which events can be tracked daily. Certain standards have to be met and these can be measured objectively, as described above, or judged subjectively. The standards may be devised to satisfy government regulations, as for instance maximum fat content in emulsion-type sausage products. They may be company standards set by the management with the aim of being above a given product quality level, or they may be set to ensure that the consumer gets a uniform and consistent level of quality in the product purchased.

Some differentiate quality assurance from quality control. Quality assurance is a strategy used by management to assure that their policy regarding quality of the product produced is enacted. Quality control is the actual carrying out of the sampling, testing and interpretation.

The corporate management must want and support a quality control program in order for it to work. The quality control manager must report directly to management, since there may be a conflict with production.

An important component of quality control is statistics—especially as related to sampling and probability. Testing may be costly, and a decision must be made about how much sampling must be done to obtain an accurate picture of the total. Reporting is often done in charts that indicate action and unacceptable limits. The charts clearly show trends and changes over time.

Sanitation may be a component of quality control and is discussed in the following section.

CLEANING AND SANITATION

Cleaning and sanitation is a subject in itself. The area and environment to be cleaned—that is the plant and the equipment used for a given food application vary enormously. Similarly, the possibilities and combinations for cleaning are rather immense. Besides shift-to-shift or day-to-day cleaning, more long-term situations like deposits and buildups must be dealt with in a special manner. Rotations of cleaning compounds and mixtures used may be recommended. Therefore, it is best to consult an expert in the area. Most plants develop a program dictated by the job to be accomplished.

A clean environment is a necessary component for producing a quality product. It is definitely associated with the topic of this book—the preservation of meat. Certain aspects are, moreover, regulated and inspected by various of the governmental agencies. Finally, there is a cost involved. So, it is a management decision if the program is to meet an average standard or to exceed it and do a genuine, quality job.

The subject of cleaning and sanitation is especially important today. More and more perishable and fragile meat products (such as modified atmosphere packed fresh meat and sous-vide type of entrees) are being introduced into the market. Food safety is a major concern, and one of the best safeguards is to have the best hygienic conditions possible for manufacture of these items. Cleaning and sanitation are key components.

A total program must take into account the building, the type of meat being handled and produced and how it flows through the building. The composition of the floors, walls and ceiling is considered. The equipment is important—not only the material from which it is constructed but also if it can be disassembled easily for cleaning. Finally, the people are important. They must have clean working clothes, hair nets, possibly gloves and must be trained. Access to toilets and hand wash stations is important.

There are four usual steps in a cleanup procedure. Major soil, such as bits of meat and smeared fat, is removed using high pressure water. Removal of soil is continued with the use of detergents, and often physical actions such as brushing. The detergents and loosened soil are then rinsed away with more water. The final step is sanitization.

Rather enormous quantities of water are used. Therefore, the quality of the water itself is important. For example, hardness reduces greatly the effectiveness of detergents and leads to surface buildups. Temperature is also important because if it is too hot the meat protein can be cooked in place leaving an undesirable surface film or buildup.

High pressure units for removing soil are effective. Another procedure is the use of foam generators that mix in detergents. The foam clings in place and therefore has more time to exert a loosening action on the soil. The hand method of brushing is also employed.

There are numerous detergents, and they fall into the general categories of alkaline and acid. They are used to penetrate and loosen the soil. They counteract hardness of water if properly formulated, but this obviously reduces their efficiency in loosening soil. Selection for effectiveness must be balanced against corrosiveness, and they must be rinsed away. The alkaline detergents used are normally mildly alkaline. Acid detergents are usually used on encrusted film and scale.

Sanitizers should be effective and rapid acting in killing microorganisms. They must also be approved for use by the regulatory agencies. Remaining soil greatly diminishes the effectiveness of sanitizers, which is the reason for doing the best possible job of removing soil. There are two general classes of sanitizers—the halogens and the surfactants.

Chlorine and iodine compounds are the halogens normally used. For example, sodium hypochlorite added to water forms hypochlorous acid that is effective and rapid acting, not only against microorganisms but also against spores. Chlorine compounds deteriorate during storage, are less effective above pH 6, are corrosive and may act as skin irritants. Iodine compounds are generally effective in the narrow pH range of 2.5 to 3.5.

The commonly used surfactants are the quaternary ammonium compounds and acid ionic compounds. They are often employed to sanitize floors, walls and aluminum equipment. They have good sanitizing activity as well as detergent properties, and may impart residual bacteriostatic activity. However, they have limited effectiveness against coliforms and gram-negative bacteria.

Steam and hot water may also be used to sanitize, with temperature and length of exposure being important factors. The most frequently employed use is water at 180°F.

The effectiveness of the cleaning and sanitizing should be routinely monitored, as part of the quality control program.

TOTAL QUALITY MANAGEMENT

The aim of total quality management is to produce a better quality product. This is accomplished by convincing everyone in the company, at all levels, to think about and participate in preventing problems in quality from occurring. This is contrasted with the usual method of detecting quality problems in the final product by inspecting it.

The production workers require special consideration. They must be trained and involved actively in the total quality management scheme. New technologies have been incorporated into production processes and numerous changes have been made. This puts much more pressure on quality control personnel, and it presents the possibility for more frequent errors or mistakes in production. Under the usual circumstances, production and quality control are often at odds. Production is working to fill their schedules, and they often view quality control as an interference that slows their production efforts. There is little effort to involve production personnel in control of quality but rather there is an adversarial situation.

Under total quality management, quality becomes everyone's responsibility and is not managed just by quality control. All workers are involved. They are instructed to think and not just do as they are told. Success of the system depends on open communication, cooperation and teamwork. The definition of quality extends beyond the usual to include safety to and acceptability by the consumer.

HAZARD ANALYSIS CRITICAL CONTROL POINTS

This technique, also known as HACCP, has also come in for a great deal of discussion recently because it affords a means for a food industry to give itself as much protection as possible against a food safety hazard.

We have discussed already that significant changes have taken place in the meat industry, and more are being put in place. For example, the interest in modified atmosphere packaging has been mentioned. It is a given fact that the consumer desires less sodium, and the industry is responding with low salt products. Consumers want fresher products and so the industry is attempting to provide them by various means and also by shipping product out more rapidly. All of this allows less time for testing of final product before it is in the hands of the consumer. The primary hurdle to some microbiological problems in meat and meat products is refrigeration. At best, refrigeration is difficult to control in distribution chains and retail outlets—unless the product's manufacturer owns the refrigeration. As discussed previously, there is little control once the product is in the consumers' hands.

The three areas for concern about safety are the raw material, the total manufacturing process and the distribution of product. Through the years the meat industry has worked out excellent procedures for preservation (safety). But now, as changes have occurred, there are questions about how well the safety is really assured. It has become obvious that the total system must be considered, from production on the farm to consumption by the consumer.

Responding to a crisis, or after the fact testing is not sufficient anymore. What is needed is a preventive system and that is what HACCP offers.

HACCP was developed by the Pillsbury Company in the late 1950's when they were asked by the National Aeronautics and Space Administration (NASA) to provide foods for the space effort. The aim was to ensure that their foods would not cause an illness due to any of numerous factors including bacteria, viruses, and chemicals contaminants. They determined, then, that they had to have control over the raw material, process, environment, personnel, storage and distribution. Besides control over and knowing as much as possible about each area, complete record keeping was essential.

The actual details of setting up and carrying out an HACCP program will not be recounted here but can be found in numerous references sources. Many training schools and programs are being offered. As a matter of general information the HACCP principles are set forth as follows:

Principle 1: Assess hazards associated with growing, harvesting, raw materials and ingredients, processing, manufacturing, distribution, marketing, preparation and consumption of food.

Principle 2: Determine the critical points required to control the identified hazards.

Principle 3: Establish the critical limits that must be met at each identified critical control point.

Principle 4: Establish procedures to monitor critical control points.

Principle 5: Establish corrective action to be taken when there is a deviation identified by monitoring of a critical control point.

Principle 6: Establish effective record keeping systems that document the HACCP plan.

Principle 7: Establish procedures for verification that the HACCP system is working correctly. Verification measures may include microbiological, physical, chemical, electronic and sensory methods.

MODELING

Activity in this area is directed at developing mathematical models to predict growth of microorganisms under various sets of conditions. The obvious advantages are that the method is truly predictive rather than retrospective, and with computer technology and a sufficient data base a large number of factors can be examined. Interest is great because as new variations of meat foods are being developed rapidly the question of food safety becomes increasingly critical. Time and cost make it essentially impossible to conduct complete retrospective studies in each case.

Of concern are the possibilities of spoilage or of production of toxin. The usual method of examination for the former has been to conduct microbiological analyses on meat samples over an extended time course. This leaves the investigator with relating time to spoilage to what happened previously in the manufacture of the meat. To study toxin production the usual procedure has been the so-called inoculated pack study. In this case a product is manufactured (inoculated, for example, with spores), and then packages are removed periodically during storage (probably being held at abuse temperature) and analyzed for presence of toxin. The results are good for that particular product under that particular set of conditions. This is an extensive and lengthy process.

Success in generating a model would permit rapid prediction of the effects of changing factors such as pH, temperature, water activity, nitrite, salt, phosphate and ascorbate in various formulation and methods of manufacture. Possible synergisms could be detected.

Data for modeling is usually generated from slurry or broth experiments, and therefore the question is raised if such results can be extended to actual meat products. Degree of initial contamination is also another factor to consider.

Continuing work in building and refining models may contribute greatly to food safety predictions in the future.

There are numerous ways to manage preservation, and many of these are parts of well-known procedures whose major aim is to control quality and safety of the final product being produced. Changing technology and meeting consumer demands may well complicate the situation. Scientific understanding coupled with commitment to excellence will always give the best result.

CHAPTER 9

SUMMARY AND CONCLUSIONS

Proper preservation of meat is important, and the consequences of not doing a good job are substantial. In the first instance, the economic considerations are real. Someone will lose money if meat spoils, and the potential consumer will be denied nutritive value as well as a satiety component. Since the possibility of meatborne disease exists, human health is also at stake.

The importance and consequence of proper preservation of meat was well-illustrated during 1993 by an outbreak of illness due to *E. coli* 0157:H7 in undercooked hamburgers in the Pacific Northwest. It resulted in numerous hospitalizations and even some deaths. The media coverage has been extensive and consumer concern is at an all-time high.

It is not at all surprising that attempts to preserve meat are recorded in early history. Enormous effort went into gathering and hunting food—in fact, it was a matter of survival. So any technique, such as chilling or cooking, that could be used to save the food, if only for another day, was sought.

Obviously, the attitude about the necessity of planning ahead changes during times of plenty and times of need. Perhaps so many problems have been claimed in our food supply during the past quarter century because the food supply has been so abundant and varied.

There have been some major improvements in how meat could be pre-served as a result of changes in civilization. Military and exploratory efforts have always had a strong influence on preservation technology because they have depended on a reliable and continuous food supply for success. Improved transportation allowed better distribution. Consider, for example, the changes in the meat industry as a result of refrigerated rail cars.

Presently, the consumer is having a major impact on how preservation methods are used as well as influencing the development of new technologies for the future. In most simplistic terms, the consumer wants appealing, healthful food that has undergone a minimum of preservation. Safety of food and nutrition-health issues are of paramount importance, and direction taken is often based more on opinion than on scientific fact. The result is a jungle of conflicting signals.

What does it take to understand meat preservation? I believe there are three subject areas of importance, and these are an appreciation of history, an understanding of how the production, slaughter and processing industries work, and a knowledge of fundamental science. Moreover, it must be realized that in biological matters, there is never an absolute or 100% certainty.

History is important because it shows an established record, and reasons why a procedure evolved as it did. While it is impossible to prove food safety for humans, by rodent feeding trials for example, assurance of reasonable safety can be gleaned from knowledge that a certain food (or food preservation procedure) has been used with no apparent ill effects over a period of several hundred years.

The production of either processed or fresh meat, when examined from farm production practices to retail presentation is a very complex matter, and normally it is segmented into various business or management components. The composition and quality of the meat is influenced by all the steps and procedures involved, and similarly, the response to preservation practices. Basically, a great possibility for control of final result exists, if an understanding of all the component steps is undertaken and control programs encompass all those steps.

Study of meat science and microbiology is necessary to understand spoilage and meatborne illness. Scientific understanding of mechanisms and processes is required in order to make improvements.

The methods of meat preservation can be grouped into categories of physical, chemical and microbiological.

Physical methods of preservation include such processes as heating, cooling and drying. They can be used in degree—such as refrigeration compared to freezing—and are effective preservation techniques. While modern or more recent refinements exist, these physical procedures are traced back to early history and are widely accepted and used at present. In addition to the preservation function, some of these methods are used to alter the properties of the product.

Chemical methods for preservation involve adding a chemical, and the most widely recognized procedure today is curing meat. Salt was recognized early as a good preservative and was used extensively. At present, most consumers have some objection to adding any chemical to food.

Preservation of meat by microbiological means takes two different forms. Fermentation is a time-tested means of preservation, and numerous examples of fermented meat products exist. Generally, the preservation properties are very good, and the products often have specific palatability characteristics as a result of the fermentation products. Another means of preservation is to take advantage of the fact that various microorganisms compete with each other.

There are two overall important points to be made. First, preservation methods are not perfect and do not give total protection. Second, the most effective preservation is achieved by a combination of methods.

The effectiveness of a preservation method is somewhat like a balancing act. It must be used to a degree or extent that achieves the preservation desired yet does not adversely alter the appearance or quality of the food. For example, heat can be applied to meat to the extent that it would kill all microorganisms, but would almost make the meat unacceptable to the consumer.

This leads to the fact that a variety of methods are often used in combination to produce the greatest preservation while keeping quality high. A fermented meat product has lowered water, contains salt and other chemicals, and it is still recommended to be kept refrigerated. The latter is to give an added measure of protection.

It is definitely evident that preservation of meat will be undertaken in the future, as an integrated system. That is, all steps from farm production to consumption by the consumer will be considered, and an attempt will be made to exert control at all steps.

Certainly, there will be new and improved technologies. But, there is always a risk that new technologies, although effective, may be rejected by consumers. So, it seems a safer and more effective route to integrate as much as possible.

This conclusion is drawn in light of the mood of the country, and is unlikely to change in the near future. There is a clear trend away from putting all effort on the sole goal of merely increasing agricultural production. Instead, more consideration is being given to efficiency of production. In addition, there are overriding concerns about environment, animal well-being, consumer issues of food safety, nutrition and health, global markets and competition and governmental regulatory activity.

Meat as a food is clearly in central focus in all of these concerns. However, let's not end on a negative note. From the historical viewpoint, meat has been a sought-after food, and has had an excellent record for wholesomeness. During the past 20 years there has been an escalating attack on meat, including arguments such as animals should not be used for human food, animal production practices harm the environment, fat in meat foods is harmful to human health, and consumption of meat is not safe from a microbiological perspective.

It has become obvious that many of these concerns about meat consumption are, in reality, only attempts by special interest groups to influence public policy. Fortunately, the consuming public is beginning to realize this influence.

There have been enormous advancements in meat preservation with the result of making available higher quality and safer meat. Further improvements can and will be made. There are proven technologies, such as irradiation, that can be adopted now. Scientific efforts will provide more new techniques for consideration in the future. Change is always questioned, but genuine improvements in technology eventually surface. They have been and will be accepted and make a positive impact on society. This has certainly been the case with meat preservation, and without a doubt, there will be further advancements and improvements in the future.

REFERENCES

Ahn, C. and M.E. Stiles. 1990. Antibacterial activity of lactic acid bacteria isolated from vacuum-packaged meats. J. Appl. Bacteriol. *69*, 302.

Anon. 1909. Food Preservation. A State of Facts in the Case, Together with References to the Authorities. Issued by The National Association for the Promotion of Public Health, New York, NY.

Anon. 1980. Dietary Salt. IFT Scientific Status Summary. Institute of Food Technologists, Chicago, IL.

Anon. 1981. Proceedings of the Red Arrow Product Company 20th Anniversary Smoke Symposium, Manitowoc, WI.

Anon. 1983. Radiation Preservation of Foods. IFT Scientific Status Summary. Institute of Food Technologists, Chicago, IL.

Anon. 1983. Danish Research Gives Edge to Vacuum Packaging of Meat. The National Provisioner. Oct. 1, p. 8.

Anon. 1986. Ascorbic acid, erythorbic acid, citric acid, sodium ascorbate and sodium citrate in fresh pork cuts. Fed. Reg. *51*, 163, 30052.

Anon. 1986. Recent Advances and Developments in the Refrigeration of Meat by Chilling. International Institute of Refrigeration, Paris, France.

Anon. 1988. Preservatives in Food. The Institute of Food Science and Technology, (UK), London, England.

Anon. 1992. IFT Freezing Technology Short Course Manual. Institute of Food Technologists, Chicago, IL.

Anon. 1992. Irradiation of poultry products. Fed. Reg. *57*, 193, 43588.

Anon. 1992. USDA tests show TSP effective in cutting chicken *Salmonella*. Food Chem. News *34*, 11.

Appert, 1920. *The Art of Preserving Animal and Vegetable Substances for Many Years*. (Translation by K.G. Bitting, Glass Container Association of America, Chicago, IL.)

Bacus, J. 1984. *Utilization of Microorganisms in Meat Processing*. Research Studies Press, Letchworth, England.

Bacus, J. and E. Bontenbal. 1991. Controlling *Listeria*. Natural sodium lactate has been recognized as an effective way to control pathogens in cured and uncured meats. Meat Poultry, June, p. 64.

Bailey, A.J. and N.D. Light. 1989. *Connective Tissue in Meat and Meat Products*. Elsevier Applied Science, New York, NY.

Ball, C.O. 1923. Thermal Process Time for Canned Food. Bulletin of National Research Council, Vol. 7, Part 1, No. 37.

Banks, J.G., R.G. Board and N.H.C. Sparks. 1986. Natural antimicrobial systems and their potential in food preservation of the future. Biotechnol. Appl. Biochem. *8*, 103.

Bauman, H.E. and R.P Wooden. 1990. Food safety management systems. Proc. Reciprocal Meat Conf. *43*, 13.

Beauchemin, M. 1990. Sous-vide technology. Proc. Reciprocal Meat Conf. *43*, 103.

Bechtel, P.J. (ed.) 1986. *Muscle Foods*. Academic Press, New York, NY.

Bendall, J.R. 1969. *Muscles, Molecules And Movement*. Heinemann Educational Books Ltd., London, England.

Benedict, R.C. 1988. Microbial attachment to meat surface. Proc. Reciprocal Meat Conf. *41*, 1.

Best, D. 1990. Preservatives—Now You See Them, Now You Don't. Prepared Foods, Aug., p. 113.

Beuchat, L.R. and D.A. Golden. 1989. Antimicrobials occurring naturally in foods. Food Technol. Jan., p. 134.

Bhunia, A.K., M.C. Johnson and B. Ray. 1988. Purification, characterization and antimicrobial spectrum of a bacteriocin produced by *Pediococcus acidilactici*. J. Appl. Bacteriol. *65*, 261.

Binkerd E.F., O.E. Kolari and C. Tracy. 1976. Pemmican. Proc. Reciprocal Meat Conf. *29*, 37.

Board, R.G. and G.W. Gould. 1991. Future prospects. In *Food Preservatives*. N.J. Russell and G.W. Gould (eds.) Blackie, London, England.

Bodiansky, S. 1991. In From the Cold. The New York Times Magazine, Dec. 22, p. 14.

Borresen, T. and J. Adler-Nissen. 1988. Food and agricultural biotechnology. Future application and needs. Status and perspectives in the processing of food and food ingredients. Biotechnol. Forum *5*, 346.

Brandly, P.J., G. Migaki and K.E. Taylor. 1966. *Meat Hygiene*, 3rd Ed. Lea & Febiger, Philadelphia, PA.

Bray, R.W. 1976. The meat story. Proc. Reciprocal Meat Conf. *29*, 61.

Briskey, E.J., R.G. Cassens and B.B. Marsh (eds.) 1970. *The Physiology and Biochemistry of Muscle as a Food*, 2. University of Wisconsin Press, Madison, WI.

Brothwell, D.R. and P. Brothwell 1969. *Food in Antiquity*. F.A. Praeger, New York, NY.

Buchanan, R.L. 1986. Processed meats as a microbial environment. Food Technol. April, p. 134.

Buehr, W. 1956. *Meat from Ranch to Table*. Williams Morrow and Co., New York, NY.

Burgess, G.H.O., C.L. Cutting, J.A. Lovern and J.J. Watterman. 1967. *Fish Handling and Processing*. Chemical Publishing Co., New York, NY.

Calm, C.E. 1904. *Sulphurous Acid and Sulphites as Food Preservatives*. Hygeian Chemical and Research Laboratory, Chicago, IL.

Carlton, H. 1941. *The Frozen Food Industry*. University of Tennessee Press, Knoxville, TN.

Cassens, R.G. 1990. *Nitrite-Cured Meat: A Food Safety Issue in Perspective*. Food & Nutrition Press, Trumbull, CT.

Cassens, R.G. 1993. Statement on March 16 before the House Agriculture Subcommittee on Department Operations and Nutrition and the Subcommittee on Livestock. Ser. No. 103-5, U.S. Govt. Printing Office, Washington, DC.

Cassens, R.G., M.L. Greaser, T. Ito and M. Lee. 1979. Reactions of nitrite in meat. Food Technol. July, p. 46.

Cheng, C.S. and G.J. Cocoma. 1989. Ascorbic Acid Use in Fresh Meat. Meat Industry Research Conf., American Meat Institute, Washington, DC.

Cliver, D.O. 1988. Virus Transmission via Foods. IFT Scientific Status Summary. Institute of Food Technologists, Chicago, IL.

Cliver, D.O. (ed.) 1990. *Foodborne Diseases*. Academic Press, New York, NY.

Copson, D.A. 1975. *Microwave Heating*, 2nd Ed. Van Nostrand Reinhold/AVI, New York, NY.

Corey, L. 1950. *Meat And Men: A Study Of Monopoly, Unionism And Food Policy*. Viking Press, New York, NY.

Czarnecki, J., J.C. Pierce Jr., J.H. White and B.W. Gardner, Jr. 1948. Meat products: A report of wartime problems. In *Subsistence Research and Development*. Vol. II. Quartermaster Food and Container Institute for the Armed Forces, Chicago, IL.

Daeschel, M.A. 1989. Antimicrobial substances from lactic acid bacteria for use as food preservatives. Food Technol. Jan., p. 164.

Dalgleish, J. McN. 1990. *Freeze-Drying for the Food Industry*. Elsevier Applied Science, New York, NY.

Davies, R., G.G. Birch and K.J. Parkers (eds.) 1976. *Intermediate Moisture Foods*. Applied Science Publishers, London, England.

Decareau, R.V. 1985. *Microwaves in the Food Processing Industry*. Academic Press, New York, NY.

Degnan, A.J., A.E. Yousef and J.B. Luchansky. 1992. Use of *Pediococcus acidilactici* to control *Listeria monocytogenes* in temperature-abused vacuum-packaged wieners. J. Food Protect. *55*, 98.

Deibel, R.H. and C.F. Niven. 1957. *Pediococcus cerevisiae*: starter culture for summer sausage. Bacteriol. Proc. p. 14.

Desrosier, N.W. 1970. *The Technology of Food Preservation*, 3rd Ed. Van Nostrand Reinhold/AVI, New York, NY.

Desrosier, N.W. and J.N. Desrosier. 1977. *The Technology of Food Preservation*, 4th Ed. Van Nostrand Reinhold/AVI, New York, NY.

Douglas, G.S. 1991. The Preservative Principles of Meat Fermentation. Technics Topics. Vol. 14, No. 2. Microlife Technics, Sarasota, FL.

Doyle, M.P. (ed.) 1989. *Foodborne Bacterial Pathogens*. Marcel Dekker, New York, NY.

Dziezak, J.D. 1986. Preservatives: antioxidants, the ultimate answer to oxidation. Food Technol. Sept., p. 94.

Dziezak, J.D. 1986. Preservatives: antimicrobial agents, a means toward product stability. Food Technol. Sept., p. 104.

Ellis, R. 1987. Process Temperature Monitoring Systems. Proc. Meat Industry. Res. Conf., American Meat Institute, Washington, DC.

Fabian, F.W. 1943. *Home Food Preservation. Salting-Canning-Drying-Freezing*. Van Nostrand Reinhold/AVI, New York, NY.

Farber, J.M. 1991. Microbiological aspects of modified-atmosphere packaging technology—A Review. J. Food Protect. *54*, 58.

Faustman, C. and R.G. Cassens. 1990. The biochemical basis for discoloration in fresh meat: a review. J. Muscle Foods *1*, 217.

Faustman, C., R.G. Cassens, D.M Schaefer, D.R. Buege and K.K. Scheller. 1989. Vitamin E supplementation of Holstein steer diets improves sirloin steak color. J. Food Sci. *54*, 485.

Faustman, C., R.G. Cassens, D.M. Schaefer, D.R. Buege, S.N. Williams and K.K. Scheller. 1989. Improvement of pigment and lipid stability in Holstein steer beef by dietary supplementation with vitamin E. J. Food Sci. *54*, 858.

Faustman, C., J.L. Johnson, R.G. Cassens and M.P. Doyle. 1990. Color reversion in beef. Fleischwirtschaft *70*, 677.

Fennema, O.R. (ed.) 1975. *Principles of Food Science*. Part II. *Physical Principles of Food Preservation*. Marcel Dekker, New York, NY.

Fennema, O.R. (ed.) 1976. *Principles of Food Science*. Part I. *Food Chemistry*. Marcel Dekker, New York, NY.

Fennema, O.R., W.D. Powrie and E.H. Marth. 1973. *Low-Temperature Preservation of Foods and Living Matter*. Marcel Dekker, New York, NY.

Fiddes, N. 1991. *Meat. A Natural Symbol*. Routledge, London, England.

Flosdorf, E.W. 1949. *Freeze-Drying*. Van Nostrand Reinhold Co., New York, NY.

Folin, O. 1914. *Preservatives and Other Chemicals in Foods: Their Use and Abuse*. Harvard University Press, Cambridge, MA.

Food and Agriculture Organization. 1990. Manual on Simple Methods of Meat Preservation. FAO publication 79. Rome, Italy.

Forrest, J.C., E.D. Aberle, H.B. Hedrick, M.D. Judge and R.A. Merkel. 1975. *Principles of Meat Science*. L.H. Freeman & Co., San Francisco, CA.

Fulks, F.T. 1991. Total quality management. Food Technol. June, p. 96.

Galloway, D.E. 1982. Packaging Processed Meats to Provide for Extended Shelf Life. The National Provisioner *186*, 40.

Garcia, H.S. and A.E. Yousef. 1990. Possibility of Inactivating Contaminating Bacteria in Ground Beef by Peroxidase System. Meat and Animal Science, Meat Processing Report. University of Wisconsin, Madison, WI.

Gerats, G.E. 1988. Quality and hygiene assurance at the meat processing line. In *Trends in Modern Meat Technology 2*. B. Krol, P.S. van Roon and J.H. Houben (eds.) Pudoc, Wageningen, The Netherlands.

Giese, J.H. 1991. Sanitation: the key to food safety and public health. Food Technol. Dec., p. 74.

Glass, M. 1991. Animal production systems in Neolithic Central Europe. British Archaeological Reports. Intl. Series 572. Oxford, England.

Goll, D.E., R.G. Taylor, J.A. Christiansen and V.F. Thompson. 1991. Role of proteinases and protein turnover in muscle growth and meat quality. Proc. Reciprocal Meat Conf. *44*, 25.

Goplen, A.O. 1979. The career of Marquis de Mores in The Bad Lands of North Dakota. State Historical Society of North Dakota, Fargo, ND.

Gould, G.W. 1992. Synergistic effects of combined processes due to disturbance of homeostasis within microorganisms. Int. Food Technol. Expo and Conf. p. 90.

Gould, W.A. 1977. *Food Quality Assurance*. Van Nostrand Reinhold/AVI, New York, NY.

Gracey, J.F. 1986. *Meat Hygiene*, 8th Ed. Bailliere Tindall, London, England.

Greaser, M.L. 1991. An overview of the muscle cell cytoskeleton. Proc. Reciprocal Mcat Conf. *44*, 1.

Greiner, E.C. 1987. Parasitology. In *The Science of Meat and Meat Products*, 3rd Ed. J.F. Price and B.S. Schweigert (eds.) Food & Nutrition Press, Trumbull, CT.

Hammes, W.P. 1990. Bacterial starter cultures in food production. Food Biotechnol. *4*, 383.

Handy, A.L. 1917. *War Food: Practical and Economical Methods of Keeping Vegetables, Fruits and Meats*. Houghton Mifflin Co., Boston, MA.

Hanson, R.E. 1990. Cooking technology. Proc. Reciprocal Meat Conf. *43*, 109.

Harlander, S.K. 1987. Biotechnology: emerging and expanding opportunities for the food industry. Nutrition Today. July/Aug., p. 21.

Harlander, S.K. 1990. The application of gene technology to improve organisms. Food Biotechnol. *4*, 515.

Hauschild, A.H.W. and B. Simonsen. 1986. Safety assessment for shelf-stable canned cured meats–an unconventional approach. Food Technol. Apr., p. 155.

Hinman, R.B. and R.B. Harris. 1947. *The Story Of Meat*. Swift and Co., Chicago, IL.

Hoagland, R., C.N. McBryde and W.C. Powick. 1917. Changes in Fresh Beef During Cold Storage Above Freezing. Bulletin No. 433, USDA, Washington, D.C.

Hollenbeck, C.M. 1977. Novel Concepts in Technology and Design of Machinery for Production and Application of Smoke in the Food Industry. Pure Appl. Chem. *49*, 1687.

Hotchkiss, J.H. and R.G. Cassens. 1987. Nitrate, Nitrite, and Nitroso Compounds in Foods. IFT Scientific Status Summary. Institute of Food Technologists, Chicago, IL.

Houben, H.H., P.S. van Roon and B. Krol. 1988. Continuous radio frequency pasteurization of sausage emulsions. In *Trends in Modern Meat Technology 2*. B. Krol, P.S. Van Roon and J.H. Houben, (eds.) Pudoc, Wageningen, The Netherlands.

Huffman, D.L. and J.C. Cordray. 1985. Use of Encapsulated Food Acids in Cured, Restructured, Hot-Processed Pork. Alkar Smoke Signals, Lodi, WI.

IAMFES. 1991. Hazard Analysis And Critical Control Point System (HACCP) Manual. International Assoc. of Milk, Food and Environmental Sanitarians, Ames, IA.

Incze, K. 1992. Raw Fermented and Dried Meat Products. Proc. Intl. Cong. Meat Sci. Technol. *37*, 829, Kulmbach, Germany.

Jay, J.M. 1992. *Modern Food Microbiology*, 4th Ed. Van Nostrand Reinhold Co., New York, NY.

Jensen, L.B. 1949. *Meat and Meat Foods. Processing and Preservation from Meat Plant to Consumer*. The Ronald Press Co., New York, NY.

Jensen, L.B. 1953. *Man's Foods: Nutrition and Environments in Food Gathering Times and Food Producing Times*. The Garrard Press, Champaign, IL.

Katsaras, K. and L. Leistner. 1988. Topographie der Bakterien in der Rohwurst. Fleischwirtschaft *68*, 1295.

Kauffman, R.G., R.G. Cassens, A. Scherer and D.L. Meeker. 1992. Variations in Pork Quality. History, Definition, Extent, Resolution. National Pork Producers Council, Des Moines, IA.

Kinsman, D.M. 1976. Meat preparation and preservation in colonial America. Proc. Reciprocal Meat Conf. *29*, 17.

Knorr, D. and P. Oxen. 1992. High Pressure Used in Combined Processes for Food Preservation. Int. Food Technol. Expo and Conf., p. 90.

Knutson, J. 1991. Meat Facts. American Meat Institute, Washington, DC.

Komarik, S.L., D.K. Tressler and L. Long. 1974. *Food Products Formulary. Volume I. Meats, Poultry, Fish, Shellfish*. Van Nostrand Reinhold/AVI, New York, NY.

Koohmaraie, M. 1988. The role of endogenous proteases in meat tenderness. Proc. Reciprocal Meat Conf. *41*, 89.

Kraft, A.A. 1986. Meat Microbiology. In *Muscle as Food.* P.J. Bechtel (ed.) Academic Press, New York, NY.

Kramlich, W.E., A.M. Pearson and F.W. Tauber. 1973. *Processed Meats*. Van Nostrand Reinhold/AVI, New York, NY.

Kyzlink, V. 1990. *Principles of Food Preservation*. Elsevier, New York, NY.

Lamkey, J.W., R.W. Leak, W.B. Tuley, D.D. Johnson and R.L. West. 1991. Assessment of sodium lactate addition to fresh pork sausage. J. Food Sci. *56*, 220.

Lanari, M.C. and N.E. Zaritzky. 1991. Effects of packaging and frozen storage temperature on beef pigments. Int. J. Food Sci. Technol. *26*, 629.

Lang, J.H. (ed.) 1984. *Larousse Gastronomique*. Crown Publishers, New York, NY.

Lawrie, R.A. 1966. *Meat Science*. Pergamon Press, New York, NY.

Lefens, M. 1987. Wilson's TenderCuts: one company's commitment to the future. Meat Proc. *26*, 58.

Leistner, L. 1990. Fermented and intermediate moisture products. Intl. Cong. Meat Sci. Technol. *36*, 842, Havana, Cuba.

Levie, A. 1970. *The Meat Handbook*, 3rd Ed. Van Nostrand Reinhold/AVI, New York, NY.

Liepe, H.V. 1983. Starter cultures in meat production. In *Food and Feed Production with Microorganisms*, G. Reed (ed.), p. 399.Verlag Chemie, Deerfield Beach, FL.

Liu, H., and B.M. Watts. 1970. Catalysts of lipid peroxidation in meats. 3. Catalysis of oxidative rancidity in meats. J. Food Sci. *35*, 596.

Lowe, B. 1946. *Experimental Cookery from the Chemical and Physical Standpoint*, 3rd Ed. John Wiley & Sons, New York, NY.

Luchansky, J.B. and M.P. Doyle. 1991. Behavior and control of *Listeria monocytogenes* in meats. *In*: Proc. Int. Cong. on *Listeria* and Food Safety, Laval, France, A. Amgar (ed.) p. 104.

Lueck, E. 1980. *Antimicrobial Food Additives: Characteristics–Uses–Effects*. Springer-Verlag, New York, NY.

Lyman, B. 1989. *A Psychology of Food*. Van Nostrand Reinhold Co., New York, NY.

Maas, M.R., K.A. Glass and M.P. Doyle. 1989. Sodium lactate delays toxin production by *Clostridium botulinum* in cook-in-bag turkey products. Appl. Environ. Microbiol. *55*, 2226.

Mermelstein, N.H. 1993. Controlling *E. coli* 0157:H7 in meat. Food Technol. April, p. 90.

Mitosumoto, M., R.G. Cassens, D.M. Schaefer, R.N. Arnold and K.K. Scheller. 1991. Improvement of color and lipid stability in beef longissimus with dietary vitamin E and vitamin C dip treatment. J. Food Sci. *56*, 1489.

Mitsumoto, M., R.G. Cassens, D.M. Schaefer and K.K. Scheller. 1991. Pigment stability improvement in beef steak by ascorbic acid application. J. Food Sci. *56*, 857.

Mitsumoto, M., C. Faustman, R.G. Cassens, R.N. Arnold, D.M. Schaefer and K.K. Scheller. 1991. Vitamins E and C improve pigment and lipid stability in ground beef. J. Food Sci. *56*, 194.

Mudgett, R.E. 1989. Microwave Food Processing. IFT Scientific Status Summary. Institute of Food Technologists, Chicago, IL.

Murray, F. 1991. Fresh meat packaging: a dynamic technology. Meat Intl. *1*, 43.

National Provisioner (The). 1981. Meat For The Multitudes. July 4 Special Issue in two volumes.

Nelson, P.E., J.V. Chambers and J.H. Rodriguez. 1987. Principles of Aseptic Processing and Packaging. The Food Processors Institute, Washington, DC.

Newsome, R.L. 1987. Perspective on Food Irradiation. IFT Scientific Status Summary. Institute of Food Technologists, Chicago, IL.

Niinivaara, F.P. 1955. Uber den Einfluss von Bakterien-Reinkulturen auf die Reifung und Umrotung der Rohwurst. Acta. Agralia. Fennica. *85*, 1.

Niven, Jr., C.F. 1987. Microbiology. In *The Science of Meat and Meat Products*, 3rd Ed. J.F. Price and B.S. Schweigert, (eds.) Food & Nutrition Press, Trumbull, CT.

Oblinger, J.L., M.P. Doyle, R.S. Flowers, J.F. Frank, R.G. Labbe, J. Lovett, J.M. Madden, R.L. Newsome, M.D. Pierson and N.R. Reddy 1988. Bacteria Associated with Foodborne Disease. IFT Scientific Status Summary. Institute of Food Technologists, Chicago, IL.

Ouali, A. 1990. Meat tenderization: possible causes and mechanisms. A review. J. Muscle Foods *1*, 129.

Ouali, A., X. Vignon and M. Bennett. 1991. Osmotic pressure changes in Postmortem Bovine Muscle: Factors of Variation and Possible Causative Agents. Int. Cong. Meat Sci. Technol. *37*, 452, Kulmbach, Germany.

Papadopoulos, L.S., R.K. Miller, G.R. Acuff, L.M. Lucia, C. Vanderzant, and H.R. Cross. 1991. Consumer and trained sensory comparisons of cooked beef top rounds treated with sodium lactate. J. Food Sci. *56*, 1141.

Parkhurst, C.R. and G.J. Mountney. 1988. *Poultry Meat and Egg Production*. Van Nostrand Reinhold Co., New York, NY.

Pearson, A.M. and R.B. Young. 1989. *Muscle and Meat Biochemistry*. Academic Press, New York, NY.

Potter, N.N. 1978. *Food Science*, 3rd Ed. Van Nostrand Reinhold/AVI, New York, NY.

Potthast, K. 1984. Liquid smoke—its use in the surface treatment of meat products. Fleischwirtschaft *64*, 328.

Price, J.R. and B.S. Schweigert (eds.) 1987. *The Science of Meat and Meat Products*, 3rd Ed. Food & Nutrition Press, Trumbull, CT.

Radouco-Thomas, C., C. Lataste-Dorolle, R. Zender, R. Busset, H.M. Meyer and R.F. Mouton. 1959. The anti-autolytic effect of epinephrine in skeletal muscle: non-additive process for preservation of meat. Food Res. *24*, 453.

Rapatz, G. and B. Luyet. 1959. On the mechanism of ice formation and propagation in muscle. Biodynamica *8*, 121.

Rector, T.M. 1925. *Scientific Preservation of Food*. John Wiley & Sons, New York, NY.

Reeves, P. 1972. *The Bacteriocins*. Springer-Verlag, New York, NY.

Rhee, K.S. and Y.A. Ziprin. 1987. Lipid oxidation in retail beef, pork and chicken muscles as affected by concentrations of heme pigments and nonheme iron and microsomal enzymic lipid peroxidation activity. J. Food Biochem. *11*, 1.

Roberts, T.A. 1989. Combinations of antimicrobials and processing methods. Food Technol. Jan., p. 156.

Rohter, L. 1992. With Plant Set to Open, Irradiated Food is Coming. New York Times, Jan. 21.

Romans, J.R. and P.T. Ziegler. 1974. *The Meat We Eat*, 10th Ed. Interstate Printers and Publishers, Danville, IL.

Root, W. and R. de Rochemont. 1976. *Eating In America*. William Morrow and Co., New York, NY.

Russell, N.J. and G.W. Gould (eds.) 1991. *Food Preservatives*. Blackie, London, England.

Rust, R. and D. Olson. 1987. Processing Workshop. A Series of Articles Published in Meat & Poultry during 1981–86. Oman Publishing, Mill Valley, CA.

Ryser, E.T. and E.H. Marth. 1991. *Listeria, Listeriosis and Food Safety*. Marcel Dekker, New York, NY.

Sacharow, S. 1983. Art and Science of Package Converting. Paper, Film and Foil Converter. Dec., p. 68.

Schillinger, U. and F. Lucke. 1990. Lactic acid bacteria as protective cultures in meat products. Fleischwirtschaft *70*, 1296.

Schneck, J.C. 1981. Liquid smoke application to cured meat. Proc. Reciprocal Meat Conf. *34*, 101.

Sebranek, J.G. 1985. Stabilizing the properties of meat products with packaging systems. Proc. Meat Industry Conf., Amer. Meat Institute, Washington, DC.

Simmons, F.J. 1961. *Eat Not This Flesh*. The University of Wisconsin Press, Madison, WI.

Sims, J.G. 1951. Meat and Meat Animals in World War II. Ag. Monograph No. 9., USDA, Washington, DC.

Sinclair, U. 1906. *The Jungle*. Signet Classics (1960), New York, NY.

Sitzmann, W. 1992. Effects of High-Voltage Pulses on Microorganisms and Food Ingredients. Int. Food Technol. Expo. and Conf., p. 172.

Skaggs, J.M. 1986. *Prime Cut. Livestock Raising and Meatpacking in the United States 1607–1983*. Texas A&M University Press, College Station, TX.

Smulders, F.J.M. 1987. Prospectives for microbial decontamination of meat and poultry by organic acids with special reference to lactic acid. In *Elimination of Pathogenic Organisms from Meat and Poultry*. F.J.M. Smulders (ed.) Elsevier, New York, NY.

Snijders, J.M.A., T.G. van Logtestijn, D.A.A. Mossel and F.J.M. Smulders. 1984. Conditions for the use of lactic acid as a decontaminant in the meat industry. Proc. Eur. Meet. Meat Res. Workers. *30*, 232. Bristol, England.

Sofos, J.N. 1989. *Sorbate Food Preservatives*, CRC Press, Boca Raton, FL.

Sokal, R.R., N.L. Oden and C. Wilson. 1991. Genetic evidence for the spread of agriculture in Europe by demic diffusion. Nature *351*, 143.

Sosnicki, A.A., R.G. Cassens, R.J. Vimini and M.L. Greaser. 1991. Histopathological and ultrastructural alterations of turkey skeletal muscle. Poultry Sci. *70*, 349.

Stansby, M.E. 1963. *Industrial Fishery Technology*. R.E. Krieger Publishing Co., Huntington, New York, NY.

Stauffer, J.E. 1988. *Quality Assurance of Food: Ingredients, Processing and Distribution*. Food & Nutrition Press, Trumbull, CT.

Stewart, K. 1977. *The Joy of Eating*. Stemmer House Publishers, Owings Mills, MD.

Tanaka, N., L. Meske, M.P. Doyle, E. Traisman, D.W. Thayer and R.W. Johnston. 1985. Plant trials of bacon made with lactic acid bacteria, sucrose and lowered sodium nitrite. J. Food Protect. *48*, 679.

Tannahill, R. 1973. *Food in History*. Stein and Day Publishers, New York, NY.

Tauber, F.W. 1976. The history of sausage. Proc. Reciprocal Meat Conf. *29*, 55.

Thompson, J.W. 1973. History of Livestock Raising in the United States, 1607–1860. Ag. History Series No. 5., USDA, Washington, DC.

Thorne, S. 1986. *The History of Food Preservation*. Parthenon Publishing, Krikby, Longdale, England.

Troller, J.A. 1983. *Sanitation In Food Processing*. Academic Press, New York, NY.

Tuma, H.J., D.H. Kropf, D.B. Erickson, D.L. Harrison, S.E. Trieb and A.D. Dayton. 1975. Frozen Meat—Its Distribution, Costs, Acceptance, and Cooking and Eating Qualities. Research Publication 166. Kansas State University, Manhattan, KS.

Vickery, J.R. 1990. *Food Science and Technology in Australia: a Review of Research Since 1900*. CSIRO, North Ryde, Australia.

Wagner, M. K. and L.J. Moberg. 1989. Present and future use of traditional antimicrobials. Food Technol. Jan. p. 143.

Wasserman, B.P., T.J. Montville and E.L. Korwek. 1988. Food Biotechnology. IFT Scientific Status Summary. Institute of Food Technologists, Chicago, IL.

World Book Encyclopedia (The). 1977. Field Enterprises Educational Corp., Chicago, IL.

Yousef, A.E., J.B. Luchansky, A.J. Degnan and M.P. Doyle. 1991. Behavior of *Listeria monocytogenes* in wiener exudates in the presence of *Pediococcus acidilactici* H or Pediocin AcH during storage at 4 or 25°C. Appl. Environ. Microbiol. *57*, 1461.

Zeuthen, P., J.C. Cheftel, C. Eriksson, M. Jul, H. Leniger, P. Linko, G. Varela and G. Vos. 1984. *Thermal Processing and Quality of Foods*. Elsevier Applied Science Publishers, New York, NY.

INDEX